BIOETHANOL: PRODUCTION, BENEFITS AND ECONOMICS

BIOETHANOL: PRODUCTION, BENEFITS AND ECONOMICS

JASON B. ERBAUM
EDITOR

Nova Science Publishers, Inc.
New York

LIBRARY OF CONGRESS CATALOGING-IN-PUBLICATION DATA

ISBN 978-1-60741-697-5

Published by Nova Science Publishers, Inc. ✦ New York

CONTENTS

PREFACE

The principle fuel used as a petrol substitute for road transport vehicles is bioethanol. Bioethanol fuel is mainly produced by the sugar fermentation process, although it can also be manufactured by the chemical process of reacting ethylene with steam. The main sources of sugar required to produce ethanol come from fuel or energy crops. These crops are grown specifically for energy use and include corn, maize and wheat crops, waste straw, willow and popular trees, sawdust, reed canary grass, cord grasses, jerusalem artichoke, myscanthus and sorghum plants. Ethanol or ethyl alcohol (C2H5OH) is a clear colourless liquid, it is biodegradable, low in toxicity and causes little environmental pollution if spilt. Ethanol burns to produce carbon dioxide and water. Ethanol is a high octane fuel and has replaced lead as an octane enhancer in petrol. By blending ethanol with gasoline we can also oxygenate the fuel mixture so it burns more completely and reduces polluting emissions. This new and important book gathers the latest research from around the globe in this promising field.

Chapter 1 - Lignocellulosic biomass refers to plant biomass which consists of cellulose, hemicelluloses, and lignin. Cellulose is a homopolymer of glucose. Hemicellulose, however, is a heteropolymer composed of several different sugars, including glucose, xylose, arabinose, galactose, mannose, and rhamnose. Cellulose and hemicellulose, when hydrolyzed into their sugars, can be converted into bioethanol fuel through well established fermentation technologies.

Bioethanol produced by fermentation of lignocellulosic biomass is currently an attractive alternative fuel to supplement the depleting stores of fossil fuels. However, sugars necessary for fermentation are trapped inside the crosslinking structure of the lignocellulose. Hence, proper pretreatment of biomass is invariably carried out prior to attempting the enzymatic hydrolysis of the polysaccharides (cellulose and hemicellulose) in the biomass. Pretreatment refers to a process that converts lignocellulosic biomass from its native form, in which it is recalcitrant to cellulase enzyme systems, into a form for which cellulose hydrolysis is much more effective.

In general, pretreatment methods can be grouped into four main categories, including physical, chemical, biological, and combined pretreatments. This paper focuses on the review of characteristics of different pretreatment technologies and envisions the future of biomass pretreatment for bioethanol production.

Chapter 2 - Wheat is the most widely grown cereal worldwide (sown on over 200 million hectares), and can be grown throughout Eastern and Western Europe. Within Europe, winter wheat is an important feedstock for bioethanol production. However, concerns over land use change associated with biofuel crops, and of competition between food and fuel have raised questions over the desirability of producing ethanol within Europe. In the UK, average wheat yields have doubled from 4 to *ca.* 8 t/ha on farm since the 1960s, and individual farmers in the most productive areas can achieve yields of >10 t/ha. Given that the main driver for biofuels within Europe is reduction in greenhouse gas (GHG) emissions, high yields are an important element in delivering environmentally sustainable bioethanol. For example increasing yields from a 7.5 to a 10 t/ha wheat crop (with no change in inputs or grain composition) would lead to a positive effect on net GHG savings, with the percentage reduction in emissions (from bioethanol relative to petrol) increasing from 48% to 62%. FAO statistics suggest that wheat yields in Eastern Europe have fallen behind those in Western Europe since the late 1980s, from *ca.* 70% of UK wheat yields at their peak, to only 40% of UK wheat yields in 2007. Therefore there is large potential for improvement in wheat yields in a number of EU countries, as well as Ukraine and CIS. Bioethanol manufacture also yields a protein rich co-product (distillers dried grains and solubles; DDGS) which can be used to replace soya imported into the EU. FAO suggest that up to 13 million hectares of former arable land could be brought back into production within Eastern Europe. If this land were to be used for biofuel production, it would avoid the need for land use change outside Europe as the DDGS produced would substitute for imported soya. The available data therefore suggest huge potential through breeding and better agronomy of wheat to supply the demands of the bioethanol industry, while having minimal impact on food supply within Europe. However, significant barriers to increasing productivity still remain, in terms of higher input (fertiliser, fuel) prices and encouraging uptake of improved wheat varieties by the industry. A balance will need to be struck between the need for higher grain prices which will stimulate investment by agribusiness, and the low grain prices needed to make wheat bioethanol production competitive with that from sugar cane.

Chapter 3 - Bioethanol production processes have been running for years and ethanol consumption has been proved to be advantageous over fossil fuels considering gas emission, especially in metropolitan areas, reduction of dependence of oil and gas imports and social aspects, such as job generation.

Despite the positive aspects cited and the rapid bioethanol expansion around the world, the concerns on its sustainability and challenges have increased. Among the drawbacks associated with bioethanol production are the high amount of fresh water consumed and the high generation of wastes, specifically the vinasse generated in the conventional processes (13 liter of vinasse/liter of ethanol produced).

One of the alternatives for overcoming these problems is the use of a continuous extractive process, which is associated to a higher productivity when compared to the traditional modes of operation adopted in the industrial sector. This system consists of three interconnected units: a fermentor, a membrane unit and a flash vessel under vacuum. The process is attractive for removing ethanol simultaneously to its production, reducing product inhibition in yeasts, resulting in high performance, productivity and non-necessity of heat exchangers.

Another point to be overcome in the industrial sector is the lack of robustness of the fermentation in the presence of fluctuations in operational conditions, which leads to changes in the kinetic parameters, with impact on yield, productivity and conversion. These changes are very common in plants of alcoholic fermentation.

Through the implementation of an extractive process associated with a robust model (under parameter re-estimation) to be used in optimization procedures and development of controllers, the performance of the fermentative processes is much improved, with gains of productivity, less waste generation and less water consumption.

In this chapter the advantages of the continuous extractive process and its kinetics are discussed through intrinsic modeling, considering substrate, ethanol and biomass inhibitions, rate of loss of viability of cells and temperature influence on the kinetics. This modeling reduces costs by allowing the investigation of suitable conditions by simulation and the development of optimizators and controllers. Besides, this chapter presents the methodology of continuous re-estimation of the kinetic parameters of existing models, which when associated to sensitivity analysis reduces the complexity of this task to obtain robust models, which are accurate even under changes in operational conditions.

Chapter 4 - The conversion of lignocellulosic biomass to fermentable sugars for low-cost fuel production represents a major challenge in global efforts to utilize renewable resources rather than fossil fuels. However, lignocellulosic materials, which represent the most abundant polysaccharides in nature, have not encountered significant utilization because of the difficulties associated with their degradation. In such kinds of biomass, the chains of carbohydrates, mainly cellulose and hemicelluloses, are embedded in a lignin matrix, which hinders their efficient degradation. Thermal, chemical, and biochemical approaches have been proposed, both individually and in combination, as promising solutions to overcome this difficulty, but the drastic conditions required by many pretreatment techniques give rise to problems when using conventional enzymes in the following saccharification step. In this context, enzymes isolated from microorganisms growing at extreme temperatures could represent an attractive solution since they are thermostable, active at high temperatures (thermophilic) and resistant to solvents and detergents. These unusual properties make them interesting candidates for the development of saccharification processes of lignocellulosic biomass. Here, author report on the hydrolysis of agricultural waste by thermostable enzymes isolated from extremophilic microorganisms. In particular, polysaccharide-degrading activities from the hyperthermophilic archaeon *Sulfolobus solfataricus*, which grows optimally at acidic pH (2.0–4.0) and high temperatures (80-87°C) and originally isolated from a solfataric field in the area of Naples, have been used to hydrolyze at high temperature agro-based raw materials such as brewer's spent grains and corn stover. The raw material has been subjected to enzyme hydrolysis directly or after preliminary pretreatments in order to compare the efficiency of the enzyme saccharification process on differently treated material. Initial results indicated that incubation of the brewer's spent grains with a combination of glycolytic enzymes from *S. solfataricus* at high temperature (80°C) and acidic pH (5.0) provided glucose, arabinose and xylose as final products.

Chapter 5 - Life cycle studies of fuel ethanol made from starch or sugar show that such ethanol tends to do not better or worse than conventional gasoline as to the emission of eutrophying and acidifying substances. Life cycle assessments are not unanimous regarding the emission of ecotoxic substances. Fuel ethanol varieties from different crops vary much in their life cycle emissions of substances that contribute to oxidizing smog. Life cycle studies of bioethanol from specified crops show diverging results regarding cumulative fossil fuel demand and the emission of greenhouse gases. If properly done and if the allocation of fuel demand to ethanol and co-products is on the basis of prices, cumulative fossil fuel demand is relatively high for ethanol currently produced from European grain or U.S. corn and relatively low for ethanol from Brazilian sugarcane, making the latter relatively conducive to energy security. The 'seed-to-wheel' emissions of greenhouse gases associated with bioethanol produced from sugar or starch are often higher than the corresponding life cycle emissions of conventional gasoline A major reduction of greenhouse emissions may be possible when cropping for bioethanol production is restricted to abandoned soils which currently sequester little carbon.

Chapter 6 - Among biofuels, bioethanol has increased in importance in many countries because it can be used directly or mixed with gasoline in combustion engines. The production of bioethanol in a fermentative process usually gives a dilute solution from which the bioethanol must be obtained in a high concentration in order to be used as biofuel. The use of bioethanol mixed with gasoline in combustion engines is associated with fewer emissions of both hydrocarbons and carbon monoxide. The production of high purity bioethanol using extractive distillation sequences with ethylene glycol or a dilute solution of NaCl as entrainers are studied in detail in terms of energy consumption and total annual costs. Conventional and complex distillation sequences are designed optimally in a computational framework implemented in Aspen Plus™ and Matlab™. The results indicate that complex distillation sequences involving thermal linking can reduce energy consumption over conventional distillation sequences using either ethylene glycol or a dilute solution of NaCl as entrainers. As a result, significant reductions in total annual cost can be obtained in the production of high purity bioethanol. This can position ethanol as a competitive biofuel when compared to gasoline.

Chapter 7 - International standards for Biodiesel (BD) characterization include two properties that are very important from the standpoint of safety in production, storage and transport of biofuels: Methanol Content (MC) and Flash Point (FP).

This chapter presents a systematic exploration of the direct relation existing between MC and FP, including electrical properties (permittivity and conductivity) that, although at present are not included in international standards for BD, provide relevant information on this subject.

Precisely known amounts of methanol (up to 2.5% V/V) were added to BD, and FP was determined by the well known Pensky-Martens method. An excellent fitting to a potential function was found between MC and FP. Moreover, the maximum allowable MC according to International Standards corresponds precisely to the minimum FP value indicated by the Standards.

Measurements of Electrical Properties (EP) carried out at different temperatures show a clear dependence between MC and the permittivity and conductivity of the samples. The results presented in this chapter show that FP and EP measurements are an interesting alternative for the verification of methanol content in biodiesel.

Chapter 8 - The world population is currently at 6.7 billion with a quarter million additional people added daily [PRB, 2007]. Energy specialists project that peak oil has already been reached and there are only about 60 years of this fuel remaining [BP, 2005]. There should be about 100 years of natural gas remaining. Slowly oil and gas supplies will decline until these fuels are exhausted. This will create a critical situation for food production because all food supply currently depends primarily on oil and gas to maintain a highly productive agriculture. The Food and Agriculture Organization reports that cereal grain production per capita has been declining continuously for the past 24 years [FAO, 1961-2008]. This is critical because grains make up 80% of world food. Food and biofuels are dependent on the same resources for production: land, water, and energy. Increased use of biofuels further damages the global environment and especially the world food system.

Chapter 9 - The worldwide depletion of fossil fuels and widespread concern over increasing atmospheric CO2 have sparked interest not only in biomaterials but also in sustainable, nonfossil-based fuels. Political instability in petroleum-producing regions has further increased the desirability of domestic fuel sources, particularly for transportation [1]. Solar and wind power are well-suited to sustainable generation of electricity, including electricity for charging vehicle batteries, but most modern vehicles are designed for liquid fuels that are best simulated by two biofuels: bioethanol and biodiesel [2]. In addition, biohydrogen is an emerging biofuel that carries energy from sunlight or organic matter, rather than petroleum, in clean-burning hydrogen (H_2), Finally, biodesulfurization of petroleum products may offer a way to mitigate some effects of petroleum use during a transition and is discussed as well.

Chapter 10 - Current ethanol production is based on corn grain or other starch or sugar sources which make up only a very small portion of plant material. With advanced bioethanol technology, ethanol can also be made from cellulose and hemicellulose (the components that give plants their structure), which make up the bulk of plant material.

Potential feedstocks for advanced bioethanol technology include corn stover (stalks and husks) and other agricultural residues, wood chips and other forestry residues, paper and other municipal wastes, food processing and other plant-derived industrial wastes, and dedicated energy crops of fast-growing trees or grasses.

Advanced technology bioethanol would supplement rather than replace grain ethanol, but the huge volume of inexpensive available feedstocks offers potential to greatly expand ethanol production and its economic and environmental benefits.

The U.S. Department of Energy National Biofuels Program is supporting research and development to lower the cost of advanced bioethanol technology, so as to make it a marketplace reality, and has set a goal to have commercial demonstration plants using agricultural residues in operation by 2005.

Chapter 11 - In light of a changing regulatory environment, concern has arisen regarding the future prospects for ethanol as a motor fuel. Ethanol is produced from biomass (mainly corn) and is mixed with gasoline to produce cleaner-burning fuel called "gasohol" or "E10."

The market for fuel ethanol, which consumes 6% of the nation's corn crop, is heavily dependent on federal subsidies and regulations. A major impetus to the use of fuel ethanol has been the exemption that it receives from the motor fuels excise tax. Ethanol is expensive relative to gasoline, but it is subject to a federal tax exemption of 5.4 cents per gallon of gasohol (or 54 cents per gallon of pure ethanol). This exemption brings the cost of pure ethanol, which is about double that of conventional gasoline and other oxygenates, within reach of the cost of competitive substances. In addition, there are other incentives such as a small ethanol producers tax credit. It has been argued that the fuel ethanol industry could scarcely survive without these incentives.

The Clean Air Act requires that ethanol or another oxygenate be mixed with gasoline in areas with excessive carbon monoxide or ozone pollution. The resulting fuels are called oxygenated gasoline (oxyfuel) and reformulated gasoline (RFG), respectively. Using oxygenates, vehicle emissions of volatile organic compounds (VOCs) have been reduced by 17%, and toxic emissions have been reduced by approximately 30%. However, there has been a push to change the oxygenate requirements for two reasons. *First*, methyl tertiary butyl ether (MTBE), the most common oxygenate, has been found to contaminate groundwater. *Second*, the characteristics of ethanol-blended RFG-along with high crude oil prices and supply disruptions-led to high Midwest gasoline prices in Summer 2000, especially in Chicago and Milwaukee. Uncertainties about future oxygenate requirements, as both federal and state governments consider changes, have raised concerns among farm and fuel ethanol industry groups and have prompted renewed congressional interest in the substance. Without the current regulatory requirements and incentives, or something comparable, much of ethanol's market would likely disappear. Expected changes to the reformulated gasoline requirements could either help or hurt the prospects for fuel ethanol (subsequently affecting the corn market), depending on the regulatory and legislative specifics. As a result, significant efforts have been launched by farm interests, the makers of fuel ethanol, agricultural states, and the manufacturers of petroleum products to shape regulatory policy and legislation.

This report provides background concerning various aspects of fuel ethanol, and a discussion of the current related policy issues.

In: Bioethanol: Production, Benefits and Economics
Editor: Jason B. Erbaum

ISBN 978-1-60741-697-5

Chapter 1

LIGNOCELLULOSIC BIOMASS PRETREATMENT FOR BIOETHANOL PRODUCTION

Yi Zheng [a,*], ***Ruihong Zhang*** [a]
[a] Biological and Agricultural Engineering Department
University of California, Davis
One Shields Avenue, Davis, CA 95616, USA

ABSTRACT

Lignocellulosic biomass refers to plant biomass which consists of cellulose, hemicelluloses, and lignin. Cellulose is a homopolymer of glucose. Hemicellulose, however, is a heteropolymer composed of several different sugars, including glucose, xylose, arabinose, galactose, mannose, and rhamnose. Cellulose and hemicellulose, when hydrolyzed into their sugars, can be converted into bioethanol fuel through well established fermentation technologies.

Bioethanol produced by fermentation of lignocellulosic biomass is currently an attractive alternative fuel to supplement the depleting stores of fossil fuels. However, sugars necessary for fermentation are trapped inside the crosslinking structure of the lignocellulose. Hence, proper pretreatment of biomass is invariably carried out prior to attempting the enzymatic hydrolysis of the polysaccharides (cellulose and hemicellulose) in the biomass. Pretreatment refers to a process that converts lignocellulosic biomass from its native form, in which it is recalcitrant to cellulase enzyme systems, into a form for which cellulose hydrolysis is much more effective.

In general, pretreatment methods can be grouped into four main categories, including physical, chemical, biological, and combined pretreatments. This paper focuses on the review of characteristics of different pretreatment technologies and envisions the future of biomass pretreatment for bioethanol production.

* Corresponding contact. Tel.: +1 530 752 8039; Fax: +1 530 752 2640. *E-mail address:* yizheng@ucdavis.edu (Y. Zheng).

INTRODUCTION

The increasing concerns about environmental protection, the increase of oil prices, the decrease of the world reserves of fossil energy carriers, and national security have motivated research over the past 30 years into renewable and domestic sources of fuels and chemicals now mostly derived from petroleum. Such concerns have attracted scientific interest to the use of bioethanol as a transportation fuel. The U.S. fuel ethanol industry represents an on-going success story for the production of renewable fuels. According to the BiofuelsDigest (2008), U.S. ethanol production capacity reached 7.5 billion gallons by the end of 2007, a 40 percent increase over 2006, and national production capacity increased to 13.3 billion gallons at 136 facilities in 2008 based on the completion of all existing projects. This industry forms an infrastructure from which future growth in cellulosic substrates utilization may occur. Demand for fuel ethanol is expected to increase. Currently practiced technologies in fuel ethanol industry are primarily based on the fermentation of glucose derived from starch and sucrose. However, the corn and sugar crops to ethanol industry draws its feedstock from a food stream and is quite mature with little possibility of process improvements (Ingram and Doran, 1995).

Lignocellulosic biomass, which includes forestry residue, agricultural residue, yard waste, wood products, animal and human wastes, etc., is a renewable resource that stores energy from sunlight in its chemical bonds (McKendry, 2002). It has great potentials to reduce the cost of fuel ethanol production because it is less expensive than starch (e.g. corn) and sugar (e.g. sugarcane and sugar beet) producing crops and available in large quantities. Lignocellulosic biomass feedstocks typically contain 50%-80% by weight carbohydrates that are polymers of 5C and 6C sugar units. Most of all these carbohydrates can be processed either chemically or biologically by breaking the chemical bonds to extract energy in the form of biofuels such as bioethanol, and as many as possible need to be converted to maximizing bioethanol production.

One of the first requirements in the utilization of lignocellulose for production of ethanol is to efficiently produce a fermentable hydrolysate rich in glucose from the cellulose present in the feedstock. Employment of enzymes for the hydrolysis of the lignocellulose is considered the prospectively most viable strategy to offer advantages over other chemical conversion routes of higher yields, minimal byproduct formation, low energy requirements, mild operating conditions, and environmentally friendly processing (Ghose and Ghosh, 1978; Himmel, 1999; Kadam et al., 1999; Mandels et al., 1974; Saha, 2000; Sheehan and Van Wyk, 2001; Wingren et al., 2005). It has long-term potential for cost reductions compared to other more established routes such as concentrated acid and two-stage dilute acid hydrolysis, even though the enzymatic route has the highest current costs (Lynd et al., 1991; Schell et al., 2003).

The physico-chemical and structural composition of indigenous lignocellulose is, however, recalcitrant to direct enzymatic hydrolysis of the cellulose and hemicellulose present in lignocellulosic biomass. Therefore, proper physico-chemical pretreatment step is necessary to render the cellulose amenable to enzymatic breakdown before the enzymatic hydrolysis step (Mosier et al., 2005a) as represented in the schematic diagram of Figure 1.

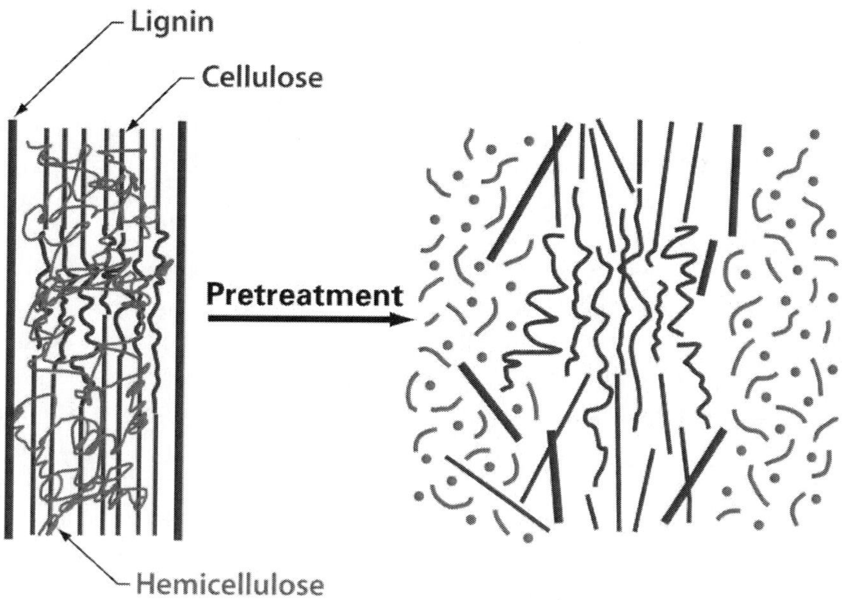

Figure 1. Schematic of pretreatment goal on lignocellulosic material (adapted from Mosier et al., 2005a).

The biomass pretreatment goal is to break the shield formed by lignin and hemicellulose, disrupt the crystalline structure of cellulose, and modifiy micro- and/or macro-structural properties of biomass. Pretreatment has been viewed as one of the most expensive processing steps in biomass-to-fermentable sugars conversion with costs as high as 30 cents/gallon ethanol produced (Mosier et al., 2005a). With the advancement of studies on pretreatment, it is believed to have great potential for improving the efficiency and lowering the cost of pretreatment (Kohlmann et al., 1995; Lee et al., 1994; Lynd et al., 1996; Mosier et al., 2003a, b).

Pretreatment techniques have been developed for various end uses of lignocellulosic biomass feedstocks. The subject of this article emphasizes the pretreatment for ethanol production and briefly reviews the aspects related to the lignocellulosic biomass pretreatment in preparation for enzymatic hydrolysis and microbial fermentation of sugars to ethanol.

OVERVIEW OF BIOETHANOL PRODUCTION FROM LIGNOCELLULOSIC BIOMASS

Production of bioethanol from lignocellulosic biomass consists of three major steps, including pretreatment, hydrolysis, and fermentation. Pretreatment is essential to alter the biomass macroscopic and microscopic size and structure as well as its submicroscopic structural and chemical composition to facilitate rapid and efficient hydrolysis of carbohydrates to fermentable sugars (Chang and Holtzapple, 2000). Hydrolysis refers to the processes that convert the carbohydrate polymers into monomeric sugars. Although a variety of hydrolysis process configurations have been studied to realize the conversion of cellulosic biomass into fermentable sugars, enzymatic hydrolysis of cellulose provides opportunities to improve the technology so that cellulosic bioethanol is competitive when compared to other liquid fuels on a large scale (Wyman, 1999). The fermentable sugars obtained from hydrolysis process could be fermented into ethanol by ethanolic fermenting microorganisms, which can be either naturally occurring or genetically modified. Cellulose in lignocellulosic biomass is usually organized into microfibrils, each measuring about 3 to 6 nm in diameter and containing up to 36 glucan chains having thousands of glucose residues. According to the degree of crystallinity, cellulose has two statuses in plant cell wall such as crystalline and paracrystalline (amorphous) cellulose. Cellulose can be hydrolytically broken down into glucose either enzymatically by cellulytic enzymes or chemically by sulfuric or other acids. Hemicellulose, a branched polymer composed of pentose (5-carbon) and hexose (6-carbon) sugars, can be hydrolyzed by hemicellulases or acids to release its component sugars, including xylose, arabinose, galactose, glucose and/or mannose. Hexoses such as glucose, galactose, and mannose are readily fermented to ethanol by many naturally occurring organisms, but the pentoses including xylose and arabinose are fermented to ethanol by few native strains, and usually at relatively low yields. Since xylose and arabinose in hemicellulose generally comprises a significant fraction of lignocellulosic biomass, especially hardwoods, agricultural residues and grasses (Wiselogel et al., 1996), it must be utilized to make economics of cellulosic ethanol processing feasible (Lynd, 1996). The development of genetic engineered microorganisms resulted in bacteria and yeasts capable of co-fermenting pentoses and hexoses into ethanol and other value-added products at high yields (Aristidou and Penttila, 2000; Bothast et al., 2002; Deanda et al., 1996; Dien et al., 1998, 2003; Ho et al., 1998, 1999; Ingram et al., 1998, 1999; Ingram and Doran, 1995; Jeffries and Shi, 1999; Mohagheghi et al., 1998; Ruohonen et al., 2006; Sues et al., 2005; Zhang et al., 1995).

With consideration of cellulase production and the process configuration of hydrolysis and fermentation, four biologically mediated events occur in the course of the production of bioethanol from lignocellulosic biomass via the technical route enzymatic hydrolysis, including cellulase production, cellulose hydrolysis, hexose fermentation, and pentose fermentation (Lynd, 1996). Process configurations proposed for the biological steps differ in the degree to which these events are integrated (Figure 2).

Enzymatic hydrolysis conducted separately from fermentation step is known as separate hydrolysis and fermentation (SHF), which involves four bioreactors. Cellulose hydrolysis performed in the presence of the hexose fermentative microorganisms with cellulase production that is carried out in a separate step by different organisms is referred to as simultaneous saccharification and fermentation (SSF). When the saccharification of both cellulose (to glucose) and hemicellulose (to xylose, arabinose, galactose, etc.) carried out in SSF with the presence of fermenting microorganisms (usually genetically engineered) which can ferment both hexose and pentose into ethanol, the SSF would become simultaneous saccharification and co-fermentation (SSCF).

Both SSF and SSCF are preferred since both hydrolysis and fermentation can be done in the same tank, resulting in lower costs (Wright et al., 1987). Consolidated bioprocessing (CBP), featuring cellulase production, cellulose and hemicellulose hydrolysis and co-fermentation in one step, is an alternative approach with outstanding potential (Lynd et al., 2005). CBP is the logical endpoint in the evolution of biomass processing technology. Although CBP is less well developed than SSF, it is expected to offer the lowest costs if limitations of current systems can be overcome.

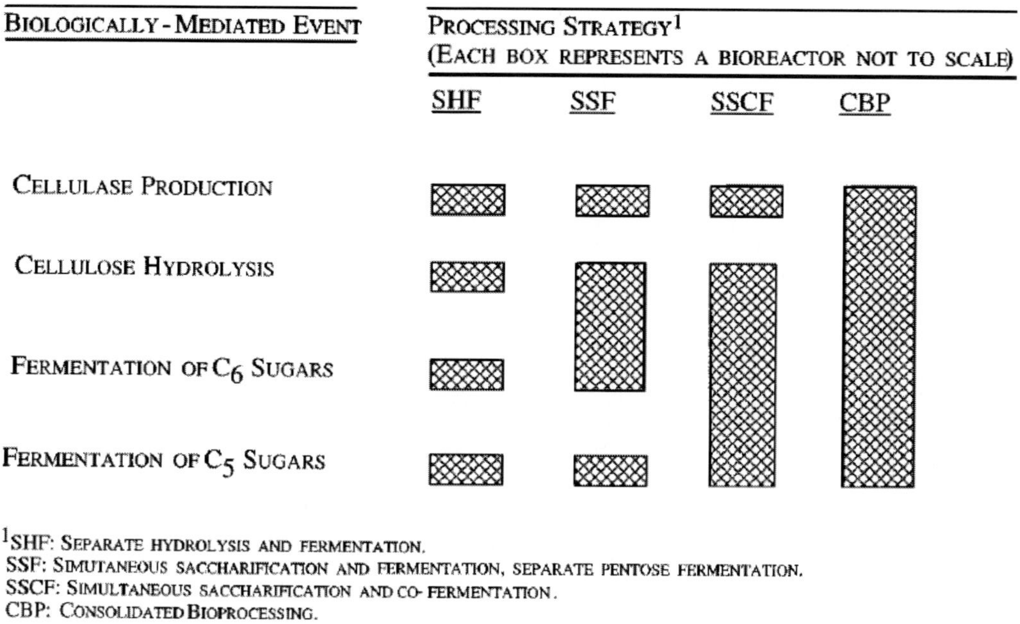

Figure 2. Consolidation of biologically mediated events in cellulosic ethanol production (adapted from Lynd, 1996).

The key difference between CBP and other biomass process configurations is that a single microbial community is employed to realize cellulase production, hydrolysis and fermentation simultaneously. This difference has several significant advantages, including no capital or operating costs for dedicated enzyme production, significantly reduced diversion of substrate for enzyme production, and compatible enzyme and fermentation systems (Lynd, 1996).

For most industrial and fuel uses, the ethanol must be purified. After fermentation, ethanol can be recovered from the fermentation broth by distillation or distillation combined with adsorption or filtration, including drying using lime or a salt, addition of an entrainer, molecular sieves, membranes, and pressure reduction (Gulati et al., 1996; Ladisch et al., 1984; Ladisch and Dyck, 1979; Onuki et al., 2008). The distillation residual solid, including lignin, ash, enzyme, organism debris, residue cellulose and hemicellulose, and other components may be recovered as solid fuel or converted to various value-added co-products (Hinman et al., 1992; Wooley et al., 1999; Wyman, 1995a, 2008).

IMPACT OF LIGNOCELLULOSIC BIOMASS COMPOSITION AND STRUCTURE ON HYDROLYSIS

Since several structural and compositional features of lignocellulosic biomass affect the enzymatic digestibility and/or fermentability, pretreatment is a necessary step to alter the structure of biomass and increase the bio-digestibility of biomass. For example, the enzymatic digestibility of native cellulose in biomass is usually lower than 20% due to its structural characteristics unless extremely excessive enzyme dose is used. The relationship between structural and compositional features of biomass and rates of enzymatic hydrolysis has been the subject of extensive research and reviews (Converse, 1993; Cowling and Kirk, 1976; Lynd et al., 2002; Mansfield et al., 1999; McMillian, 1994), but is still unclear. The structural and compositional factors of biomass commonly considered as rate-impacting factors include cellulose crystallinity, specific surface area of cellulose, degree of polymerization, cellulose protection by lignin, hemicellulose sheathing, and degree of hemicellulose acetylation (Chang and Holtzapple, 2000; Hsu, 1996; Hsu et al., 1980; Kim and Holtzapple, 2005, 2006; Rydholm, 1965; Wenzel, 1970; Yang and Wyman, 2004). However, opinions diverge as to each factor's relative contribution to the native biomass recalcitrance to enzymatic hydrolysis. The relationship between structural and compositional factors reflects the complexity of the lignocellulosic biomass systems. The variability of these characteristics explains the change of enzymatic digestibility among different sources of biomass. It also indicates the fact that the enzymatic digestibility of biomass is substrate- and pretreatment method-specific. Therefore, the relative significances of the structural and compositional barriers to enzymatic hydrolysis are different from each other depending on substrate material and various pretreatment techniques, which means that when one barrier is reduced/eliminated, another may become limiting. In principal, an effective pretreatment method or a combination of multiple pretreatments is expected to disrupt multiple barriers so that hydrolytic enzymes can penetrate and perform hydrolysis on pretreated biomass more easily. The effects of different barriers on the enzymatic hydrolysis are fully detailed as following.

Cellulose Crystallinity

Cellulose is the most abundant biomaterial on earth. Each cellulose molecule is a linear polymer of glucose residues. Depending on the degree of hydrogen bonding within and between cellulose molecules, Fan et al. (1980a, 1981a) and Lee et al. (1983) postulated a model for cellulose structure containing crystalline and amorphous fractions. Fan et al. (1982) estimated the crystalline portion of cellulose to be 50-90% of total. The high degree of hydrogen bonding that occurs among the sugar subunits within and between cellulose chains forms a 3-D lattice-like structure. The highly ordered, water-insoluble nature of crystalline cellulose makes access and hydrolysis of the cellulose chains difficult for the aqueous solutions of enzymes. Amorphous cellulose lacks this high degree of hydrogen bonding, thus rendering it a structure that is less ordered and more enzymatically digestible. The crystallinity of cellulose has often been thought of as one of big challenges to efficient hydrolysis and also one of indicators of substrate reactivity. Typically, cellulose hydrolysis rates mediated by fungal cellulases are 3-30 times faster for amorphous cellulase as compared with crystalline cellulose (Lynd et al., 2002; Teeri, 1997). Tsao et al. (1978) showed that the crystallinity weakens digestibility of the pure crystalline cellulose. Using a grass, *Miscanthus sinensis*, Yoshida et al. (2008) showed 4- and 3-fold decrease in enzymatic hydrolysis rate and yield, respectively with the increase of cellulose crystallinity, although they also found a strong correlation between digestibility and lignin, substrate site and hemicellulose. They suggested that lignin is the most significant resistance factor against the enzymatic hydrolysis of *M. sinensis*. When studying the effect of cellulose crystallinity on the enzymatic hydrolysis of biomass, Koullas et al. (1990) used ball milling to reduce the crystallinity of cellulose and achieved higher digestibility. Liu and Chen (2006) pretreated wheat straw using ionic liquid to improve its enzymatic digestibility and they indicated that the hydrolysis rates improvement was attributed to the decrease of the absolute crystallinity degree of cellulose, the polymerization degrees of cellulose and hemicellulose, and the increase of its reaction accessibility. Additional studies also found a positive correlation between crystallinity and hydrolysis and claimed that the amorphous part of cellulose was hydrolyzed first, leaving the more recalcitrant crystalline part unhydrolyzed (Gharpuray et al., 1983a; Sasaki, et al., 1979; Sinitsyn et al., 1991). However, several investigations revealed that there is few correlations between crystallinity and digestibility (Gardner et al., 1999; Han and Callihan, 1974; Mansfield et al., 1997; Nahzad et al., 1995; Ramos et al., 1993; Tanahashi et al., 1989); cellulose crystallinity of alkaline pretreated wheat straw (Koullas et al., 1993) and acid pretreated hardwood (Grethlein, 1985) unchanged or increased, even though the enzymatic digestibility of cellulose increased following pretreatment; Puri (1984) found the digestibility of bagasse pretreated by four different methods is governed by particle size, surface area, and degree of polymerization not crystallinity; lowest crystallinity achieved by ball attrition did not show the highest digestibility (Rivers and Emert, 1987); while digesting newsprint, Rivers and Emert (1988) even found that amorphous cellulose is less digestible than crystalline cellulose. Focusing on the change of substrate crystallinity during enzymatic hydrolysis, several studies (Lenze et al., 1990; Mansfield et al., 1997; Nahzad et al., 1995; Ohmine et al., 1983; Puls and Wood, 1991; Ramos et al.,1993; Schurz et al., 1985; Sinitsyn et al., 1989) did not find increased crystallinity, which is supposed to increase over the course of cellulose hydrolysis as a result of preferential reaction of amorphous cellulose (Betrabet and Paralikar, 1977; Ooshima et al., 1983). The possible reasons for the contradictory conclusions on the role of cellulose crystallinity can be: (1) the effect of crystallinity is substrate- and/or pretreatment-specific; (2) the measurement technique of crystallinity isn't so accurate (Converse, 1993; Lenze et al., 1990; Weimer et al., 1995); (3) crystallinity alone is insufficient to stop significant hydrolysis if sufficient enzyme is used (Caulfield and Moore, 1974; Howell and Stuck, 1975; Lee and Fan, 1982; Zhang and Lynd, 2004); (4) the effect of cellulose crystallinity was overlapped (Pan et al., 2007). Considering both the uncertainty of crystallinity measurement technologies as well as conflicting results on the change of crystallinity during pretreatment and hydrolysis, it would be arbitrary to conclude that cellulose crystallinity is a key determinant of the rate of enzymatic hydrolysis (Lynd et al., 2002; Mansfield et al., 1999). The development and application of improved methods would be useful for future research to more definitively resolve the role of crystallinity in impacting hydrolysis. In analyzing both crystallinity as well as other cellulose physical properties, it is important to distinguish correlation from cause and effect.

Specific Surface Area

Cellulose is an insoluble and porous substrate so that the hydrolysis of cellulose is a heterogeneously catalytic process. The hydrolytic enzymes such as cellulase enzymes must bind to the surface of substrate particles before the initiation of hydrolysis. The 3D structure of cellulose particles in combination with the size and shape of the cellulase enzymes under consideration determine whether cellulose surface are or are not accessible to enzymatic attack. The specific surface area is considered one of most important cellulose's structural features affecting the rate and extent of hydrolysis (Bertran and Dale, 1985; Fan et al., 1980a, 1981a; Gharpuray et al., 1983a; Grethlein, 1985; Grous et al., 1986; Lee et al., 1982; Tokuyasu et al., 2008). Surface area of cellulose can be divided into exterior and interior surface areas (Cowling and Brown, 1969; Ladisch et al., 1981; Stone et al., 1969). The exterior surface area is affected mainly by the shape and size of fiber meaning smaller size with greater amount of surface area, while the interior surface area depends on the capillary structure of the cellulose fibers and the size of the penetrating reactant (Chandra et al., 2007). The interior surface area of cellulose is generally 1-2 orders higher than the exterior surface area (Chang et al., 1981). When we discuss the surface area, we have to know the common methods used to qualify the surface area. The interior surface area can be measured by small angle X-ray scattering, water vapor sorption, size exclusion, N_2 adsorption on a dry sample (Brunauer-Emmett-Teller, BET method), and mercury porosimetry (Fan et al., 1980a; Gharpuray et al., 1983a; Grethlein, 1985; Hernadi, 1984; Neuman and Walker, 1992; Stone et al., 1969). Solute exclusion has become the preferred method for measuring specific surface area because of its aqueous environment – the same enzymatic hydrolysis environment – as opposed to the dry environment during N_2 adsorption which had ever been widely used to measure specific surface area, but may promote the collapse or shrinkage of the pore structure (Gharpuray et al., 1983a; Grethlein, 1985; Stone et al., 1969). In addition, N_2 can access to pores and cavities on fiber surface that cellulase can't enter, because the molecule of N_2 is much smaller than cellulase. As a result, the basis to use surface area measured by BET method is limited, and thus the determinant role of surface area is not conclusive (Mansfield et al., 1999). In general, techniques for measuring interior surface do not estimate exterior area (Converse, 1993). External surface area can be estimated with a particle counter by using a microscopy (Gilkes et al., 1992; Henrissat et al., 1988; Lee and Fan, 1982; Marshall and Sixsmith, 1974; Reinikainen et al., 1995; Weimer et al., 1990; White and Brown, 1981). Unlike cellulose crystallinity and DP, the conclusions for the effect of specific surface area on the enzymatic hydrolysis is consistently in literature, which means higher surface area causes greater digestibility of cellulosic biomass.

In general, two methods have been used to increase the specific surface area of cellulosic biomass such as size reduction and swelling process to increase the pore volume. Reduction of particle size is a very common pretreatment method to increase the enzymatic hydrolysis of cellulosic biomass, and it was found that smaller particle size has a greater amount of surface area leading to higher hydrolysis rate and extent (Grethlein, 1986; Kim et al., 1992; Mandels et al., 1971; Mooney et al., 1999; Rotter et al., 2008; Teramoto et al., 2008; Zeng et al., 2007). However, Zhang and Lynd (2004) argued that the results achieved by size reduction may be due to reasons other than increased exterior area, perhaps decreasing mass transfer resistance, since exterior surface is thought to be a small fraction of overall surface area for most substrates. Several studies supported these arguments. Although it is apparent that particle size has a significant effect on cellulose hydrolyzability, it has also been shown that the increase of pore volume has a greater effect on the improvement of hydrolysis rate and extent by cellulases than does a decrease in particle size (Esteghlalian et al., 2001; Grethlein, 1985, 1986; Grethlein et al., 1984; Grous et al., 1986; Ishizawa et al., 2007; Jeoh et al., 2007; Liu and Chen, 2006; Mooney et al., 1998; Nahzad et al., 1995; Oksanen et al., 2000; Shevchenko et al., 2000; Stone et al., 1969; Sun and Chen, 2008; Tanaka, et al., 1988; Thompson et al., 1992; Walker and Wilson, 1991; Weimer and Weston, 1985; Wong et al., 1988). Through extensive research, Grethlein et al. (1985) also found linear correlations between the initial hydrolysis rate of pretreated biomass and the pore size. They continued that the rate-limiting pore size for the hydrolysis of lignocellulosic substrates was 5.1 nm, similar to the size of *T. reesei* cellulase components. Comparing the hydrolysis rates on various sources of model cellulosic substrates, Fierobe et al. (2002) concluded that accessibility of cellulose is an important factor in determining the hydrolysis rate. It is likely that the interior surface area is more important than exterior surface area affecting the enzymatic hydrolysis of cellulosic biomass. Although the specific surface area of the substrate provided by decreased particle size and increased pore volume plays a significant role in facilitating hydrolysis by cellulases, it might not a sole factor that affects the digestibility of native cellulosic biomass (Han and Ciegler, 1983; Ishizawa et al., 2007; Laureano et al., 2005; Liu and Chen, 2006; Millett et al., 1976; Myerly et al., 1981; Tsao et al., 1978). It may correlate to other substrate factors, including crystallinity, DP, lignin protection, or combination of those factors.

Although extensive results have been achieved on the specific surface area, two critical questions are not addressed yet. The first question is how to accurately measure the enzyme-adsorbing surface area? The second question is how to specifically indentify the enzyme-specific surface area? Since some enzymes adsorb onto cellulose nonspecifically, not all enzyme-adsorbing surface areas is specific surface area. The surface area measured by solute exclusion is usually overestimated, because the solute molecular is much smaller than the cellulase enzymes. Therefore, efforts should be taken to find more accurate techniques to measure the specific surface area of cellulosic biomass in order to answer these two question.

Degree of Polymerization

There is correlation between cellulose solubility and degree of polymerization (DP). Cellulose solubility decreased rapidly with increasing DP due to intermolecular hydrogen bonds. Cellodextrins with the range of DP from 2-6 and from 7-30 are soluble in water and somewhat soluble in hot water, respectively (Pereira et al., 1988; Schmid et al., 1988; Zhang and Lynd, 2003). When DP is higher than 30, the glucan responds the polymer cellulose in its structure and properties (Klemm et al., 1998). The DP of cellulosic biomass determines the relative abundance of terminal and interior β-glucosidic bonds, and of substrates for exo-acting and endo-acting enzymes, respectively (Zhang and Lynd, 2004). As well known, endoglucanases act on interior portions of the cellulose chain to shorten the length of cellulose polymer and thus rapidly decrease DP (Kleman-Leyer et al., 1992, 1994; Selby, 1961; Srisodsuk et al., 1998; Whitaker, 1957; Wood and McCrae, 1978), while increasing the number of reducing chain ends without generating appreciable solubilization (Irwin et al., 1993; Kruus et al., 1995; Reverbel-Leroy et al., 1997). On the contrary, exoglucanases act on the reducing chain ends, and then reduce DP only incrementally (Kleman-Leyer et al., 1992, 1996; Srisodsuk et al., 1998). Wood (1975) found that exoglucanases have a marked preference for substrates with lower DP. As a result, exoglucanases and endoglucanases should work synergistically in order to solubilize cellulose during hydrolysis. Therefore, the relative proportion of exo- and endo- activities of hydrolytic enzymes and the properties of cellulose determines the change of DP during the hydrolysis of cellulosic substrates. Although few evidences showed relationship between the creation rate of reducing chain end by endoglucanase and substrate DP, DP reduction by pretreatment was found to improve the enzymatic hydrolysis of cellulosic biomass (Mansfield et al., 1997; Zhang and Lynd, 2004). It is usually difficult to evaluate the effects of DP exclusively, because altered DP can be associated with crystallinity or accessible surface area. However, there have been a few studies investigating the effects of the DP of cellulosic biomass on their hydrolysis by cellulases. Some investigations concluded that DP was less important to enzymatic hydrolysis (Engstrom et al., 2006; Mansfield et al., 1999; Nahzad et al., 1995; Sinitsyn et al., 1991). Other researchers, however, claimed that substrates with lower DP were hydrolyzed more quickly and extensively than those with higher DP (Martinez et al., 1997; Pan et al., 2008; Puri, 1984). For example, Puri (1984) found the rate and extent of saccharification was governed by particle size, surface area, and degree of polymerization, but not crystallinity, when using four different methods (carbon dioxide explosion, alkaline explosion, Ozone treatment, and sodium chlorite) to pretreat Bagasse. It was believed that higher DP formed stronger networks by more extensive inter- and intramolecular hydrogen bonding, therefore limiting the access of enzymes to cellulose and thus diminishing the digestibility (Puri, 1984). In addition, the decrease of DP increased the number of cellulose chain ends available to exoglucanases, thus generating high hydrolysis rate and yield (Martinez et al., 1997; Valjamae et al., 2001; Zhang and Lynd, 2004). The contradictory conclusions result from, to great extent, the accuracy of methodologies of DP measurement. In other words, the effect of DP was overlapped by other factors, such as crystallinity, surface area, lignin removal, etc. Therefore, in investigating the substrate physical factors that affect hydrolysis, cares must be taken to the continual evolution of analytical techniques such as thermoporosimetry (Maloney et al., 2003) and high resolution fiber quality analysis (OpTest Equipment, 1999), which may be capable of dealing with the diversity of pretreated lignocellulosic substrates with high accuracy.

Cellulose Protection by Lignin

Cellulose exists within a matrix of other polymers, primarily hemicellulose and lignin. Lignin is a complex, highly cross-linked aromatic polymer composed of phenylpropane units (Adler, 1977; Sarkanen, 1975). It is covalently linked to hemicellulose, thus stabilizing the mature cell wall. It is generally thought of as “glue” that binds cellulose and hemicellulose, reinforcing both strength and durability of lignocellulosic structure. As such, lignin clearly protects cellulose from environmental exposure. Actually, lignin is probably the most recognized inhibitor of cellulase, thus causing native cellulose resistance to enzymatic hydrolysis. Both the amount and type of lignin depend on the source of lignocellulosic biomass (Chandra et al., 2007; Sjostrom, 1981; Zheng et al., 2007). In addition, the presence of lignin carbohydrate complexes (LLCs) that is composed of lignin linked to carbohydrates through bonds such as ester, ether or ketal is another factor restricting hydrolysis processes (Eriksson and Lindgren, 1977; Lai, 1991). Therefore, many pretreatment methods have been explored to expand the pore structure by breaking LLCs and/or reduce the lignin content of cellulosic substrate to improve the enzymatic digestibility of substrate. However, most of the studies have been concerned with decreasing lignin content by altering pretreatment conditions or methods, and few investigations have been performed to link changes of lignin structure to the enzymatic hydrolysis of cellulosic substrate, probably because the effect of the modification of lignin structure was overlapped by other factors such as hemicellulose removal and decrease of DP. Alkaline (Chundawat et al., 2007) and organosolv pretreatment

(Araque et al., 2008) methods remove significant amount of lignin, while a large portion of lignin remains intact in the solid phase after steam explosion (Ohgren et al., 2007) or dilute acid pretreatment (Zheng et al., 2007). Whichever such pretreatments is used, the digestibility of pretreated biomass is always increased. Therefore, the pretreatment which improves the cellulose or hemicellulose yield ought to undoubtedly affect the lignin content because lignin is interlinked among cellulose and hemicellulose (Bura, 2004; Pan et al., 2006). The modification of lignin surfaces is believed to mitigate the inhibition of lignin on cellulase by reducing the non-productive adsorption of cellulase to lignin (Zheng et al., 2008).

There is always positive correlation between lignin removal and enzymatic hydrolysis of cellulosic biomass (Avgerinos and Wang, 1983; Draude et al., 2001; Gharpuray et al., 1983b; Gould 1984, 1985; Kim and Holtzapple, 2005, 2006; Koullas et al., 1993; Mooney et al., 1998; Ohgren et al., 2007; Pan et al., 2004, 2005; Shimizu, 1981; Yang et al., 2002; Zhu et al., 2008). Pan et al. (2005) reported 30% improvement of hydrolysis with only 7% delignification of steam-pretreated Douglas-fir substrate. The correlation levels off at 40-50% delignification, based on the results of Gharpuray et al. (1983b), Gould (1984, 1985), and Zhu et al. (2008). However, Kim and Holtzapple (2005, 2006), Pan et al. (2004), Shimizu (1981), and Yang et al. (2002) showed the enzymatic digestibility of pretreated cellulose increased up to 90% delignification.

Overall, it has been demonstrated that the existence of lignin in cellulosic biomass is a significant barrier to enzymatic hydrolysis. However, the lignin removal is usually accompanied by hemicellulose solubilization more or less (Grethlein, 1985). When performing low-temperature alkaline pretreatment, Avgerinos and Wang (1983) achieved 70% lignin removal and only 5% pentose loss. On the contrary, Ohgren et al. (2007) used steam-pretreatment to approach 70% enzymatic hydrolysis and found as high as 40% hemicellulose solubilization with 50% delignification under the pretreatment conditions of T=190 °C, t=5min, and SO_2 concentration=3%. Several other researchers also found delignification processes resulted in significant hemicellulose solubilization (Carrasco and Roy, 1992; Chum et al., 1987; Gould, 1984; Lora and Pye, 1992; Thompson et al., 1990; Yoon et al., 1994). Unlike lignin, the recovery of hemicellulose component sugars is quite sensitive to changes in processing conditions, and hemicellulose hydrolysis can potentially be used to fortify recovered sugars to increase ethanol yields in subsequent fermentation and make cellulosic ethanol more economical (Bura, 2004). As a result, the digestibility improvement of pretreated cellulosic biomass may not be attributed to the effect of delignification alone.

Hemicellulose Sheathing

Hemicellulose is a major lignocellulose component. It is heterogeneous, branched sugar polymer composed of mostly pentoses (xylose and arabinose), some hexoses (mannose, glucose and galactose), and acetylated sugars. Within the plant cell wall structure, hemicellulose is thought to surround the cellulose-fibrils like a sheath resulting in a reduced accessibility of the cellulose-fibrils (Grethlein, 1985). In addition, the covalent link between hemicellulose and lignin also forms a strong "shield" providing cellulose recalcitrance to enzyme attack and other bio-degradations. It has been widely accepted that hemicellulose acts similarly to lignin as a barrier within the lignocellulosic matrix prohibiting access of enzymes to cellulose (Grethlein and Converse, 1991; Tanaka et al., 1988; Varga et al., 2004). Hemicellulose blocks the contact of cellulase with cellulose by adsorbing enzyme and by physically blocking access of the cellulase to the cellulose surface (Yoon, 1998). Therefore, the hydrolysis of the hemicelluloses either by chemical or by biochemical method is essential to facilitate complete cellulose degradation. Several pretreatment techniques such as SO_2 steam-explosion and dilute acid pretreatment have been extensively studied to degrade hemicellulose to improve cellulose digestibility.

Zhu et al. (2005a), who pretreated corn stover using dilute sulfuric acid in a percolation reactor, indicated that the digestibility of pretreated corn stover was related to the extent of xylan removal. By using sulfuric and phosphoric acid pretreatments, Um et al. (2003) concluded that cellulose digestibility of corn stover was investigated as a function of hemicellulose removal and it was more directly related to hemicellulose removal than to delignification. Studies of dilute acid pretreatment on wheat straw and aspen showed that hemicellulose removal substantially enhanced cellulose digestion despite of high lignin content (Grohmann et al., 1984, 1985, 1986). Knappert et al. (1980, 1981), studying dilute sulfuric acid pretreatment of several different biomass, including poplar, oak, newsprint, etc., found that increased glucose yields were due to hemicellulose removal by acid pretreatment. In the case of steam-pretreated substrates, strong correlations were also found between increasing hemicellulose removal and enhanced cellulose hydrolysis. Increasing solubilization of hemicellulose during pretreatment has facilitated subsequent cellulose hydrolysis by cellulases (Boussaid et al., 2000; Fernandez et al., 2001; Mcdonald and Clark, 1992; Ruiz et al., 2008; Vlasenko et al., 1997).

Although hemicellulose removal alone increases the surface area and pore volume and makes cellulose more accessible to cellulase, it has been reported that lignin subjected to high-temperature acid pretreatments may undergo partial chemically melting and re-condensed as an altered lignin polymer, which may reduce its detrimental role in affecting digestibility of cellulose (Grohmann et al., 1985; Torget et al., 1991; Yang and Wyman, 2004). Furthermore, Grohmann et al. (1986) indicated that at least 50% of hemicellulose should be removed to significantly increase cellulose digestibility. And also, some earlier thoughts don't seem to particularly consider the presence of hemicellulose as a factor in native cellulose digestibility (Fan et al., 1982; Millett et al., 1976; Tsao et al., 1978). The effect of hemicellulose is only part of the "hydrolysis puzzle", as it is likely that changes in the primary components of lignocellulose also have significant effects on the physical and chemical characteristics of the substrate (Chandra et al., 2007). Therefore, research that attributes increasing cellulose digestibility to hemicellulose solubilization could not consider the effect of hemicellulose solubilization alone on the enzymatic hydrolysis of biomass.

Most studies on pretreatment technology have been centering on maximizing sugar yield from cellulose by optimizing pretreatment conditions. However, these optimal conditions differ from those that target maximum hemicellulose and lignin yields (Heitz et al., 1991). High pretreatment severity usually causes high hemicellulose solubilization and high cellulose digestibility, but it does not mean the pretreatment severity can increase constantly to achieve complete cellulose conversion because hemicellulose recovery will decrease due to hemicellulose degradation after pretreatment severity exceed a "threshold". Hemicellulose degradation products such as furfural and hydroxymethyl furfural (HMF) were well proved to inhibit subsequent fermentation (Clark and Mackie, 1984). Since hemicellulose component sugars usually take up to 50% of available carbohydrate, recovery and utilization of these sugars for the production of fuels and chemicals can significantly help improve the economic feasibility of biorefinery. As a result, maximizing the total sugar yield while minimizing inhibitor production must be balanced against the negative effect of hemicellulose degradation on the improvement of enzymatic digestibility of the substrate. In other words, pretreatment conditions may need to be frequently tailored considering the compromise between removing the lignin and hemicellulose components from cellulose while concomitantly maximizing the recovery of all the available carbohydrates (Chandra et al., 2007). Different pretreatment method approaches the recovery of hemicellulose in a distinct manner, as various pretreatment methods all have varying effects on the hemicellulose fraction.

Degree of Hemicellulose Acetylation

Compared with other compositional and structural features of lignocellulosic biomass, the degree of hemicellulose acetylation has received relatively little attention. Xylan is the major hemicellulose in the lignocellulosic biomass, especially in hardwoods and grasses (Timell, 1967; Wilkie, 1979). Xylan backbone in native plant cell wall is extensively acetylated (Browning, 1967; Holtzapple, 1993; Sjostrom, 1985; Timell, 1967). Several studies have shown that deacetylation greatly enhanced biomass digestibility through increased swellability, thereby increasing the hydrolysis rate (Grohmann et al., 1989; Kim and Holtzapple, 2006; Kong et al., 1992; Linden et a., 1994; Teixeira et al., 2000). Working on wheat straw and aspen, Grohmann et al. (1989) found that enzymatic digestibility of cellulose was improved by 2-3 times due to the deacetylation of xylan. This effect leveled off about 75% deacetylation, except for which other factors, such as the presence of lignin, became limiting. Similarily, Kong et al. (1992) studies aspen, but he found that the cellulose digestibility increased with the deacetylation up to 100%. He also revealed that the cellulose digestibility further increased by delignification after the effect of deacetylation leveled off. However, Zhu et al. (2008), studying poplar wood, found that acetyl content had a lesser effect on digestibility than lignin content or biomass crystallinity, and the small effect of acetyl content on digestibility may result from the low acetyl content in raw biomass (ca. 3%). The discrepancy on the effect of deacetylation could be because the acetyl group is much more sensitive to hydrolysis than other factors.

Because compositional and structural features are closely associated, in the other word, the change in one feature may also lead to changes in the others, thus they are generally called as structural features (Zhu et al., 2008). Therefore, it is difficult even impossible to exclusively study the effects of specific features on the enzymatic digestibility of lignocellulosic biomass. Without considering the action mode of pretreatment, an ideal pretreatment technique should be able to maximize the recovery of available carbohydrate such as cellulose and hemicellulose while minimizing the degradation of sugars and the generation of potential inhibitors. To summarize this section, the correlations between structural features and enzymatic digestibility of cellulosic biomass is shown as Table 1. Again, there are contradictory conclusions on the effects of several features. The possible reasons would be: (1) those features are sensitive to pretreatment; (2) close associations among features make studies on the individual effect difficult even impossible; (3) the effects of different features are usually overlapped. More research need to be done for the development of more advanced analytical techniques in order to quantify the features more specifically.

Table 1. Correlation between structural features and enzymatic digestibility of cellulosic biomass (Adapted from Zhu et al., 2008)

Structural features	Relationship between features and enzymatic digestibility
Cellulose crystallinity	Negative or no correlation
Degree of polymerization	Negative or no correlation
Specific surface area	Positive
Cellulose protection by lignin	Negative
Hemicellulose sheathing	Negative
Degree of hemicellulose acetylation	Negative or no correlation

General Considerations on Pretreatment

The general purpose of pretreatment is to open the structure of the lignocellulosic biomass, thus making it more accessible than native materials to the hydrolytic enzymes and/or generating soluble oligo- and monosaccharides. This goal can be accomplished in different ways such as changing biomass composition and modifying biomass structure. Various pretreatment technologies have been extensively studied to process diverse biomass for cellulosic ethanol production. However, none of those can be declared a "winner" because each pretreatment has its inherent advantages and disadvantages. According to the National Research Council (1999), an effective pretreatment is characterized by several criteria: avoiding size reduction, preserving hemicellulose fractions, limiting formation of inhibitors due to degradation products, minimizing energy input, and being cost-effective. Except for these criteria, several other factors are also needed to be considered, including recovery of high value-added co-products (e.g., lignin and protein), pretreatment catalyst, catalyst recycling, and waste treatment. When comparing various pretreatment options, all the mentioned criteria should be comprehensively considered as a basis. In addition, pretreatment results must be weighed against their impact on the ease of operation and cost of the downstream processes and the trade-off between several costs, including operating costs, capital costs, and biomass costs (Delgenes et al., 1996; Ladisch et al., 1983; Lynd et al., 1996; Palmqvist and Hahn-Hagerdal, 2000; Wyman, 1995b, 1996, 1999).

Pretreatment Processes

Based on the application and type of pretreatment catalyst (liquid and steam water are not considered a catalyst in this article), pretreatment techniques have generally been divided into three distinct categories, including physical, chemical, and biological processes. Combination pretreatment by incorporating two or more pretreatment techniques from the same or different categories is also common (Hsu, 1996; McMillan, 1994), but it is not grouped as an individual pretreatment category. Depending on the biomass feedstock, the mechanical size reduction step is required for most pretreatment processes. According to the review by Hsu (1996), the acceptable largest particle size of biomass for pretreatment is that of commercial wood chips, or approximately in the range of 1 to 3 cm for length and x 0.5- to 1 -cm thickness and people usually use particles passing a 3 mm or smaller rejection screen. Such size reduction step is not categorized as a distinct pretreatment process in this article. Various pretreatment processes are briefly reviewed as following based on their characteristics and action modes.

Physical Pretreatment

As reflected by the name, physical pretreatment does not employ chemical reagents. The previously studied physical pretreatment typically include mechanical comminution, pyrolysis, irradiation, uncatalyzed steam explosion (autohydrolysis), and hydrothermolysis.

Mechanical comminution: biomass is comminuted by various chipping, grinding and milling. The milling can be further detailed into hammer- and ball- milling (wet, dry, and vibratory rod/ball milling) (Horton et al., 1980; Gracheck et al., 1981; Millet et al., 1976; Rivers and Emert, 1987; Yoshida et al., 2008), compression milling (Ryu et al., 1982; Tassinari, 1980; 1982; Tassinari and Macy, 1977), ball milling/beating (Schurz, 1986), agitation bead milling (Horton et al., 1980), pan milling (Zhang et al., 2007a), and other types of milling (fluid energy milling, colloid milling, two roll milling) (Fan et al., 1982; Mandels et al., 1974; Shimizu, 1980). In addition, attrition (Rivers and Emert, 1987; Ryu and Lee, 1983) and disk refining (Han et al., 1978; Horton et al., 1980; Schurz, 1986) were also used for pretreatment. The biomass had ever been pretreated by simultaneous ball milling/attrition and enzymatic hydrolysis (Kelsey and Shafizadeh, 1980; Mais et al., 2002; Neilson et al., 1982; Ryu and Lee, 1983). Vibratory ball milling was found to be more effective than ordinary ball milling on the improvement of biomass digestibility when used to pretreat spruce and aspen chips (Millet et al., 1976). Among all the mechanical comminution techniques, the compression milling is the only process that has been tested in production-scale (Hsu, 1996).

Mechanical comminution primarily disrupts cellulose crystallinity, decreases DP, and increases the specific surface area of cellulosic biomass by breaking down the biomass into smaller particles and rendering the substrate more amenable to subsequent enzymatic hydrolysis. Usually, the mechanical comminution is time-consuming, energy-intensive and expensive. Cadoche and Lopez (1989) compared the power requirement of mechanical comminution of hardwood, straw and corn stover based on the final particle size. Based on the results from Cadoche and Lopez (1989), Sun and Cheng (2002) made a summarization as Table 2. Furthermore, mechanical comminution is much less effective than phsico-chemical and chemical pretreatments since it does not result in lignin removal, which has been proved to significantly restrict accessibility of cellulose and inhibit cellulases. Basically, it is seldom used nowadays as a pretreatment method individually.

Pyrolysis: To date, pyrolysis has received extensively studies as a thermochemical conversion technique to produce biofuels. In the late 1970s, it had been investigated as a pretreatment process to increase the susceptibility of cellulosic material to hydrolysis (Fan et al., 1987). When temperature is higher than 300°C, cellulose rapidly decomposes to produce gas and a small amount of tar residue–char. (Bradbury et al., 1979; Kilzer and Briodo, 1965;

Shafizadeh, 1977; Shafizadeh et al., 1978). At intermediate temperatures, decomposition becomes slower and fewer volatile products are generated.

Table 2 Energy requirement of mechanical comminution of biomass with different size reduction (Adapted from Sun and Cheng, 2002)*

Lignocellulosic materials	Final size (mm)	Energy consumption (kWh/ton)	
		Knife mill	Hammer mill
Hardwood	1.60	130	130
	2.54	80	120
	3.2	50	115
	6.35	25	95
Straw	1.60	7.5	42
	2.54	6.4	29
Corn stover	1.60	NA	14
	3.20	20	9.6
	6.35	15	NA
	9.5	3.2	NA

*NA: Not available.

Using vacuum pyrolysis, Shafizadeh (1977) achieved 70-78% tar yield in the temperature range of 375 to 425°C, and he used 0.5 M sulfuric acid to hydrolyze tar at 97°C for 2.5 h to attain 80-85% reducing sugar yield with over 50% glucose of cellulose. Although the presence of oxygen can benefit the pyrolytic depolymerization, the formation of unwanted by-products of oxidation and dehydration is also accelerated. In order to retard the occurrence of oxidation and dehydration during pyrolysis and increase the depolymerization, Fan et al. (1980b) conducted pyrolysis in air or inert atmosphere at relatively low temperature of 170°C and observed a negligible change in the crystallinity index and surface area of Solka Floc. However, they obtained a significant increase in the hydrolysis rate for Solka Floc treated in helium atmosphere due to depolymerization of cellulose. It is likely that low temperature plus low oxygen concentration can promote the depolymerization of cellulose and inhibit the generation of undesirable by-products. $ZnCl_2$ and Na_2CO_3 have been reported to cause decomposition of pure cellulose at much lower temperatures (Emsley and Stevens, 1994; Shafizadeh and Bradbury, 1979; Shafizadeh and Lai, 1975). Pyrolysis could be a potential pretreatment method; however, the residues need to be detoxified before enzymatic hydrolysis and fermentation since pyrolysis by-products such as levulinic acid and HMF are strong inhibitors to enzymatic hydrolysis or fermentation when biological processes are used for cellulosic biomass conversion (Purwadi, 2006). As a pretreatment technology, pyrolysis receives few research interests currently.

Irradiation: Digestibility of cellulosic biomass has been enhanced by the use of high energy irradiation methods, including γ-ray (Beardmore et al., 1980; Blouin and Arthur, 1958; 1960; Dunlap et al., 1976; Fan et al., 1980b; Kumakura and Kaetsu, 1979; Pritchard et al., 1962; Yang et al., 2008; Youssef and Aziz, 1999), ultrasound (Imai et al., 2004; Mahinpour and Sarbolouki, 1998; Nitayavardhana et al.,2008; Wojciak and Pekarovicova, 2001), electron beam (Bak et al., 2009; Horton et al., 1980; Khan et al., 1986, 1987; Kumakura and Kaetsu, 1978, 1984; Lawton et al., 1951; Shin and Sung, 2008), pulsed electrical field (Zheng and Delwiche, 2008), UV (Dunlap and Chiang, 1980), and microwave heating (Kitchaiya et al., 2003; Ma et al, 2009; Ooshima et al., 1984; Saha et al., 2008; Zhu et al., 2005b). The action mode behind the irradiation pretreatment could be one or more changes of features of cellulosic biomass, including increase of specific surface, decrease of the degrees of polymerization and crystallinity of cellulose, hydrolysis of hemicellulose, and partial depolymerization of lignin. Based on the literature, these high energy irradiation methods were not cost-effective. They are usually slow, energy-intensive, or prohibitively expensive (Chang et al., 1981; Fan et al., 1982; Han, et al., 1978; Lin et al., 1981). Irradiation appears to be strongly substrate-specific; for example, the digestion of aspen carbohydrate is essentially complete after an electron dosage of 108 rads, while spruce is only 14% digestible at this dosage (Dunlap and Chiang, 1980). Millett et al. (1976) have concluded that irradiation techniques lack commercial appeal based on an estimation of overall cost. Therefore, most irradiation methods were not used any more, except for microwave and electron beam. Currently, microwave and electron beam are not used alone. They should be used with the presence of some chemicals such as oxygen, nitrate and nitrite salts (Blouin and Arthur, 1960; Dunlap et al., 1976; Dunlap and Chiang, 1980). Or they can be used to aid the alkaline or acid pretreatment, which is discussed in the section of combination pretreatment.

Uncatalyzed steam-explosion/autohydrolysis: Uncatalyzed steam-explosion is one of the most common pretreatment methods for lignocellulosic biomass. Extensive research has been done on this method (Chandra et al., 2007; Hsu, 1996; McMillan, 1994; Saddler et al., 1993). It is one of only a very limited number of cost-effective pretreatment technologies that have been advanced to pilot scale demonstrate and commercialized application. Commercial equipment is available. Mason (1926) and Delong (1981) used the uncatalyzed steam-explosion to hydrolyze hemicellulose for commercially produce fiberboard and other products. In this method, biomass particles are rapidly heated by high-pressure saturated steam for a period time to promote the hemicellulose hydrolysis without use of any chemicals. This process is terminated by swift release of pressure, which renders the biomass undergo an explosive decompression (Abatzoglou et al., 1992; Avellar and Glasser, 1998; Brownell and Saddler, 1984; Glasser and Wright, 1998; Heitz et al., 1991; Ramos et al.,1992; Saddler et al., 1982, 1983; Wyman et al., 1993). During the pretreatment, the hemicellulose is often hydrolyzed by organic acids such as acetic acids and other acids formed from acetyl or other functional groups, released from biomass. In addition, water, itself, also possesses certain acid properties at high temperature, which further catalyze hemicellulose hydrolysis (Baugh et al., 1988a, b; Weil et al., 1997). The uncatalyzed steam-explosion is also termed autohydrolysis in the literature. Therefore, the degradations of glucose and/or xylose might happen during uncatalyzed steam-explosion (Cantarella et al., 2004; Garcia-Aparicio et al., 2006). The action mode of uncatalyzed steam-explosion is similar to that of acid aid-chemical pretreatment, except that, during steam-explosion, the biomass is heated rapidly by steam so that much less moisture exists in the reactor resulting in much more concentrated sugars in comparison. The key factors for uncatalyzed steam-explosion are treatment time, temperature, particle size and moisture content (Ballesteros et al., 2002; Duff and Murray, 1996; Negro et al., 2003a). Uncatalyzed steam-explosion is typically conducted at a temperature of 160-270°C for several seconds to a few minutes before pretreated contents are discharged into a vessel for cooling. Lower temperature and longer residence time are more desirable (Wright,

1988), and the use of very small particles in steam explosion would not be favorable in optimizing the effectiveness of the process improving economy (Ballesteros et al., 2002).

The major physico-chemical changes of lignocellulosic biomass during the uncatalyed steam-explosion are attributed to the hemicellulose removal and lignin transformation. These changes improve the digestibility of biomass to enzymes. Enzymatic digestibility of pretreated poplar chips reached 90%, compared to only 15% hydrolysis of untreated chips (Grous et al., 1986). Additional, the rapid thermal expansion opens up the biomass particle structure leading to the reduction of particle size and increased pore volume (Michalowicz et al., 1991); however, it was believed less important in enhancing the digestibility of steam exploded lignocellulosic biomass (Biermann et al., 1984; Brownell et al., 1986). Although uncatalyzed steam-explosion can effectively improve the digestibility of the pretreated cellulose residue, it suffers from low hemicellulose sugar yield (Excoffier et al., 1991; heitz et al., 1991; Wright, 1988). Therefore, it is still a problematic option for long-term ethanol production (Hsu. 1996).

In addition, there are two pretreatments similar to high pressure steaming, including moist-heat expansion (extrusion) and dry-heat expansion (popping) (Fan et al., 1982). They received much less research attention than the steam-explosion. Data reported by Han and Callihan (1974) showed that extrusion pretreatment is ineffective in increasing the digestibilities of rice straw and sugarcane bagasse after 90s pretreatment. However, using the extrusion technique to pretreat newspaper for its acid hydrolysis, Brenner et al. (1977) and (Fan et al., 1982) obtained remarkable results and suggested that extrusion may be a promising pretreatment method for acid hydrolysis.

Liquid hot water pretreatment (LHW): In liquid hot water pretreatment, pressure is utilized to maintain water in the liquid state at elevated temperatures (Brandon et al., 2008; Dien et al., 2006; Negro et al., 2003b; Rogalinski et al., 2008, Walch et al., 1992). Biomass undergoes high temperature cooking in water with high pressure. This technique was invented in 1930 for wood processing not for ethanol production (Aronovsky and Gortner, 1930). It has been currently employed as a pretreatment method to improve the ethanol production from lignocellulosic biomass. As proceeding, LHW has been named hydrothermolysis (Bobleter, 1994; Bobleter and Concin, 1979; Bobleter et al., 1976, 1981; Bonn et al., 1983), uncatalyzed solvolysis (Mok and Antal, 1992, 1994), aquasolv (Allen et al., 1994, 1996; Kubikova et al., 2000), and aqueous or steam/aqueous fractionation (Bouchard et al., 1990, 1991). LHW pretreatment has been reported to have the potential to enhance cellulose digestibility, sugar extraction, and pentose recovery, with the advantage of producing prehydrolyzates containing little or no inhibitor of sugar fermentation (Van Walsum et al., 1996). It has been shown to remove up to 80% of the hemicellulose and to enhance the enzymatic digestibility of pretreated material in herbaceous feedstocks, such as corn fiber (Allen et al., 2001) and sugarcane bagasse (Allen et al., 1996; Laser et al., 2002). Perez et al. (2007) used LHW to pretreat wheat straw and obtained maximum hemicellulose-derived sugar recovery of 53% and enzymatic hydrolysis yield of 96%. They found that the optimal values of hemicellulose-derived sugar recovery and enzymatic hydrolysis yield associated with the LHW pretreatment of wheat straw fell within different temperature and pretreatment time intervals, suggesting optimization of these responses on an individual basis. Perez et al. (2008) continued to optimize process variables (temperature and residence time) in LHW pretreatment of wheat straw and achieved 80% and 91% xylose recovery and enzymatic hydrolysis, respectively, based on individual optimization. The potential ethanol yield would be 88% of the theoretical ethanol yield based sugars in raw material. The LHW is usually conducted in three types of reactors (Figure 3), including co-current (Beery et al., 2000; Mosier et al., 2003a,b; Weil et al., 1998a), countercurrent (Thomsen et al., 2008), and flow through (Liu and Wyman, 2004). LHW reduces the need for neutralization of liquid streams and conditioning chemicals since acid is not added (Mosier et al., 2005a; Negro et al., 2003b; Perez et al., 2007). Additionally, biomass size reduction is not needed because the particles are broken apart during pretreatment; therefore, LHW appears attractive (Kohlman et al., 1995; Weil et al., 1997). The sugar-enriched liquid prehydrolyzates can be fermented to ethanol (Lynd et al., 1996; Mosier et al., 2003a, b; van Walsum et al., 1996).

Although LHW is similar to the uncatalyzed steam-explosion, except for the state of water and the use of explosion, the action mechanism is different due to different moisture levels in two pretreatments. During LHW, the cleavage of *O*-acetyl and uronic acid substitutions from hemicellulose produce acetic acid and other organic acids, which help catalyze the hydrolysis of polysaccharide such as hemicellulose into soluble oligosaccharides first, and then monomeric sugars. Under acidic condition, these monomeric sugars are subsequently partially degraded to aldehydes such as furfural and 5-HMF, which are inhibitors to fermenting microorganisms.

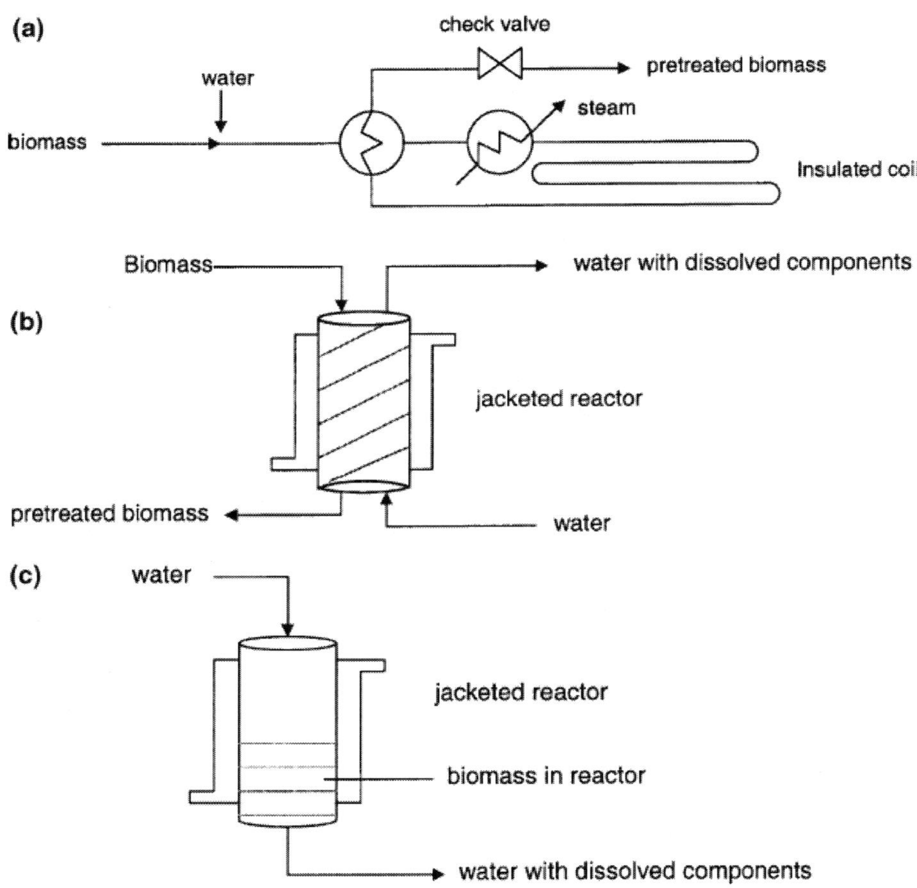

Figure 3. Schematic illustrations of liquid hot water pretreatment: (a) Co-current; (b) Counter-current; (c) Flow through (adapted from Mosier et al., 2005a).

Additionally, hot water has an unusually high dielectric constant and enables to dissolve almost all hemicellulose and a certain amount of lignin, depending on the temperature. Therefore, hot water, itself, plays a role like an acid to hydrolyze hemicellulose to release sugars and acids (Antal, 1996). In order to control the pH of the liquid hot water between 5 and 7, some bases such as KOH is usually added into LHW pretreatment process with its role to maintain the pH not as a catalyst for alkaline pretreatment. This method is termed pH controlled hot water pretreatment, which is discussed below in detail.

Chemical Pretreatment

The pretreatment which has some chemicals (gases or liquids) involved is defined as chemical pretreatment. Chemical pretreatments were originally developed and have been extensively used in the paper industry for delignification of cellulosic materials to produce high quality paper products. They have been considered, however, quite sever and expensive to be used for pretreatment of lignocellulosic biomass for ethanol production. The possibility of developing effective and inexpensive pretreatment techniques by modifying the pulping processes deserves consideration (Fan et al., 1982). Most chemical pretreatments that have been studied to date have had the primary goal of improving the enzymatic digestibility of cellulose by removing lignin and/or hemicellulose, and to a lesser degree decreasing the DP and crystallinity of the cellulosic component. Chemical pretreatments have received by far the most attention among all pretreatment category. Various chemical pretreatment methods are discussed under several sub-categories, namely, catalyzed steam-explosion, alkaline, acid, organosolv, ammonia fiber/freeze explosion, pH-controlled liquid hot water, ionic liquids, wet oxidation, and others.

Catalyzed steam-explosion: Catalyzed steam-explosion is very similar to uncatalyzed steam-explosion on their action modes, except that some chemicals, including SO_2, H_2SO_4, CO_2, oxalic acid, etc. are used as catalysts to impregnate the biomass prior to steam-explosion. It is recognized as one of the most cost-effective pretreatment processes. SO_2- and H_2SO_4-impregated steam-explosions have been tested in pilot-scale (Fein et al., 1991; Ropare et al., 1992). Compared with uncatalyzed steam-explosion, it has more complete hemicellulose removal leading to more increased enzymatic hydrolysis of biomass with less generation of inhibitory compounds (Morjanoff and Gray, 1987). A number of studies on the catalyzed steam-explosion pretreatment have been reported (Boussaid et al., 1999; Clark and Mackie, 1987; De Bari et al., 2007; Mackie et al., 1985; Tengborg et al., 1998). SO_2 appears more appealing than H_2SO_4 in steam-explosion since the former requires milder and much less expensive reactor material, generates less gypsum, yield more xylose, and produces more digestible substrate with higher fermentability. Although several researchers (Puri and Mamers, 1983; Zheng, et al., 1995, 1998) found CO_2-explosion (with or without steam) is effective, Dale and Moreira (1982) observed that CO_2-explosion was less effective than ammonia fiber/freeze explosion (AFEX). Hohlberg et al. (1989) and Mamers and Menz (1984) also concluded that CO_2 was less effective than SO_2 in steam-explosion. SO_2-impregantaed steam-explosion is the only known pretreatment technique that can effectively make softwoods more digestible (Millett et al., 1976; Stenberg et al., 1999), even though it was found less effective for softwood than hardwood and herbaceous biomass (Clark and

Mackie, 1987). However, SO_2 is highly toxic and may present negative safety, heath and environmental impacts. Again, the catalyzed steam explosion also generates some inhibitors derived from the degradation of carbohydrates. Certain detoxification strategies might be needed if inhibitors are detrimental to downstream processes. Additional limitations of steam explosion include destruction of a portion xylan faction and incomplete disruption of the lignin-carbohydrate matrix (Mackie et al., 1985).

Alkaline pretreatment: Alkaline pretreatment is one of major pretreatment technologies received numerous studies. It employs several bases, including sodium hydroxide (Abdi et al., 2000; Carrillo et al., 2005; Farid et al., 1983; Pinto and Kamden, 1996; Silverstein et al., 2007), calcium hydroxide (lime) (Chang et al., 1997, 1998, 2001; Kaar and Holtzapple, 2000; Kim and Holtzapple, 2005), potassium hydroxide (Chang and Holtzapple, 2000), aqueous ammonia (Foster et al., 2001; Kim et al., 2003; Mes-hartree et al., 1988; Yoon et al., 1995), ammonia hydroxide (Prior and Day, 2008), and sodium hydroxide in combination with hydrogen peroxide or others (Azzam, 1989; Mishima et al, 2006; Patel and Bhatt, 1992; Saha and Cotta, 2006, 2007). Alkaline pretreatment is basically a delignification process, in which a significant amount of hemicellulose is generally solubilized as well. The action mechanism of alkaline pretreatment is believed to be saponification of intermolecular ester bonds crosslinking xylan hemicelluloses and other components, for example, lignin and other hemicellulose. In addition, alkali pretreatments also remove acetyl and various uronic acid substitutions on hemicellulose that enhance the accessibility of the enzyme to the hemicellulose and cellulose surface (Chang and Holtzapple, 2000). Dilute NaOH treatment of lignocellulosic materials caused swelling, leading to decreased DP, increased in internal surface area, decreased crystallinity, disruption of the lignin structure, and separation of structural linkages between lignin and carbohydrates (Fan et al., 1987). The effectiveness of alkaline pretreatment varies, depending on the substrate and treatment conditions. In general, alkaline pretreatment is more effective on hardwood, herbaceous crops, and agricultural residues with low lignin content than softwood with high lignin content (Feist et al. 1970; Bjerre et al., 1996; Millet et al., 1976). Millet et al. (1976) observed that the digestibility of NaOH-treated hardwood increased from 14 to 55% with the decrease of lignin content from 24-55% to 20%, however, slight effect of dilute NaOH pretreatment was found for softwoods with lignin content greater than 26%. Kim and Holtzapple (2005) used lime to pretreat corn stover and obtained maximum lignin removal of 87.5% at 55°C for 4 weeks with aeration. Using lime pretreatment at ambient conditions for up to 192 h, Playne (1984) enhanced the enzyme digestibility of the sugarcane bagasse from 20% to 72%. He also concluded that lime would be the choice chemical based on the chemical cost. In our lab, we are currently working on the application of potassium hydroxide for rice straw pretreatment for ethanol production and found that potassium hydroxide pretreatment significantly improved the biodigestibility of rice straw.

In addition, aqueous ammonia was also employed to delignify lignocellulosic biomass. Detroy et al. (1980) found slight increased enzymatic digestibility of wheat straw pretreated by aqueous ammonia. A process named ammonia recycled percolation (ARP) is a common reactor configuration for aqueous ammonia pretreatment. Yoon et al. (1994) used ARP to significantly enhance the poplar wood digestibility. The delignification efficiencies of corn cob and switchgrass were improved by 80% and 85%, respectively by ARP pretreatment (Iyer et al., 1996). ARP was somewhat less efficient in pretreatment of softwood-based pulp mill sludge (Kim et al., 2000). A recent study achieved the enzymatic digestibility of the ARP treated corn stover of 90% with an enzyme loading of 10 FPU/g-glucan (Kim et al., 2002). Ammonia sulfite had been used for conventional pulping process, which was modified by Clarke and Dyer (1973) to pretreat lignocellulosics for animal feed. However, it was never used for cellulosic ethanol process.

In comparison with other pretreatment technologies, alkali pretreatment usually uses lower temperatures and pressures, even ambient conditions. Pretreatment time, however, is recorded in terms of hours or days much longer than other pretreatment processes. A significant disadvantage of alkaline pretreatment is the conversion of alkali into irrecoverable salts and/or the incorporation of salts into the biomass during the pretreatment reactions so that the treatment of a large amount of salts becomes a big issue for alkaline pretreatment. Compared to acid pretreatment, alkaline pretreatment is much gentler on the rector material and the current cost of alkaline is much cheaper than sulfuric acid of which price was much cheater than alkali before 2005. Nevertheless, the concentration of alkali is usually comparable to or higher than that of acid. Eggeman and Elander (2005) have attempted to compare the process and economics for ethanol production employing several pretreatments and conclude that little differentiation between the projected economic performances of the pretreatment options.

Acid pretreatment: Acid pretreatment method is derived from the concentrated acid hydrolysis such as concentrated H_2SO_4 and HCl hydrolysis which has been used to treat lignocellulosic materials (Goldstein et al., 1983; Harris and Kline, 1949; Kobayashi, et al., 1960; Sakai, 1965; Vedernikov et al., 1991). The concentrated acid hydrolysis had been temporarily commercialized in the World War II. Concentrated acid has been initially applied to remove hemicellulose either in combination with hydrolysis of cellulose to glucose or prior to acid hydrolysis of cellulose. Although it is powerful and effective for cellulose hydrolysis, concentrated acid is toxic, corrosive and hazardous and require reactors that need expensive construction material resistant to corrosion. Additionally, the concentrated acid must be recovered and recycled after hydrolysis to render the process economically feasible (Sivers and Zacchi, 1995). Therefore, it has phased out gradually. Alternatively, dilute acid pretreatment has received numerous research interests. It has been successfully developed for pretreatment of lignocellulosic biomass. Several different acids, including dilute sulfuric acid (Ballesteros et al., 2008; Martin et al., 2007a; Marzialetti, et al., 2008; Nguyen et al., 1999; Sun and Cheng, 2005; Um et al., 2003; Zheng et al., 2007; Zhu et al., 2004), dilute nitric acid (Brink, 1993, 1994), dilute hydrochloric acid (Israilides et al., 1978; Jimenez and Bonilla, 1993; Herrera et al., 2003; Mehlberg, 1977; Mehlberg and Tsao, 1979), dilute phosphoric acid (Israilides et al., 1978; Um et al., 2003; Vazquez et al., 2007), and peracetic acid (Chang and Holtzapple, 2000; Fan et al., 1981b; Teixeira et al., 1999a, b, 2000; Zhao et al., 2007, 2008a, b), have been reported in the literature. The action mode of dilute acid is to solubilize hemicellulose and remain lignin and cellulose intact so that the enzymatic digestibility of cellulose is enhanced. The oligomeric hemicellulosic saccharide could be completely hydrolyzed into monosaccharides by adjusting pretreatment conditions, but also the sugar degradation products will be generated during oligomer hydrolysis. The major advantage of dilute acid pretreatment over steam-explosion is significantly higher xylose yield. Using batch dilute sulfuric acid pretreatment process, xylose yield was shown to approach 80-90% of the theoretical yield (Grohmann et al., 1985, 1986; Grohmann and Torget, 1992; Torget et al., 1990, 1991, 1992; Torget and Hsu, 1994).

Dilute acid pretreatment has been applied to a wide range of feedstocks, including softwood (Kim, 2005; Marzialetti et al., 2008; Nguyen et al., 1999, 2000; Zheng et al., 2007), hardwood (Carrasco et al., 1994; Pinto and Kamden, 1996; Torget and Hsu, 1994), herbaceous crops (Neureiter et al., 2004; Zheng et al., 2007), agricultural residues (Karimi et al., 2006; Martin et al., 2007a; Saha et al., 2005), wastepaper (Capekmenard et al., 1992; Wayma et al., 1993), municipal solid waste (Li et al., 2007; Zheng et al., 2007), etc. Dilute acid pretreatment performed very well on most biomass species. Of all acid-based pretreatment methods, sulfuric acid has been most extensively studied since it is inexpensive and effective. Currently, the price of sulfuric acid has skyrocketed so that the economically feasibility of dilute acid pretreatment might need to be reconsidered. Two primary types of dilute acid pretreatment processes have been used, including high temperature (T>160°C), continuous-flow process for low solids loading (5-10%) (Brennan et al., 1986; Converse et al., 1989), and low temperature (T<160°C), batch process for high solids loading (10-40%) (Cahela et al., 1983; Esteghlalian et al., 1997).

Dilute sulfuric acid has some important disadvantages: (1) corrosion that mandates expensive materials of construction; (2) acidic prehydrolyzates must be neutralized before the sugars proceed to fermentation; (3) gypsum has problematic reverse solubility characteristics when neutralized with inexpensive calcium hydroxide; (4) formation of degradation products and release of natural biomass fermentation inhibitors are other characteristics of acid pretreatment; (5) disposal of neutralization salts; (6) long hydrolysis time and high enzyme loading are usually need to achieve relatively high hydrolysis yield which translate into added cost; (7) biomass particle size reduction is necessary. For example, grinding of the cellulose to 1 mm accounts for 33% of the power requirements of the entire process. Although dilute acid pretreatment can significantly improve the cellulose hydrolysis, its cost is usually higher than some physical and chemical pretreatment processes such as steam explosion and AFEX.

Organosolv: The organosolv process is a delignification process, with varying simultaneous hemicellulose solubilization. In this process, an organic or aqueous organic solvent mixture with or without an acid (HCl, H_2SO_4, oxalic, salicylic, and acetylsalicylic acid, etc.) or alkali (NaOH) catalysts is used to break the internal lignin and hemicellulose bonds (Bennani et al., 1991; Chum et al., 1990; Myerly et al., 1981; Pan et al., 2007, 2008; Sun and Cheng, 2002; Taherzadeh and Karimi, 2008). Usually, the presence of catalyst can increase the solubilization of hemicellulose and the digestibility of substrate is also further enhanced (Chum et al., 1988). The organic solvents used in the organosolv process include methanol, ethanol, acetone, ethylene glycol, triethylene glycol, tetrahydrofurfuryl alcohol, glycerol, aqueous phenol, aqueous n-butanol (Chum et al., 1988; Fan et al., 1982; Taherzadeh and Karimi, 2008; Thring et al., 1990). Of all tested solvents, ethanol is the most used in the reported research. The usual operation temperature of organosolv is in the range of 150-200°C. Both Aziz and Sarkanen (1989) and Sarkanen (1980) found that the addition of catalyst was unnecessary for satisfactory delignification when temperature was higher than 185°C. Organic solvents are costly and their use requires high-pressure equipments due to their high volatility. The used solvents should be recovered by e.g. evaporation and condensation, and recycled to reduce the operational costs of the process. Removal of solvents from the pretreated biomass is necessary because the residual solvents may be inhibitors to the growth of organisms, enzymatic hydrolysis, and fermentation. For economic reasons, the use of low-molecular-weight alcohols such as ethanol and methanol has been favored (Chum et al., 1985; Sidiras and Koukios, 2004). The main advantage of the use of solvents over other chemical pretreatments is that relatively pure, low molecular-weight lignin is recovered as a by-product (Katzen et al., 1995). Based on the review by Aziz and Sarkanen (1989), the organosolv process is too complex and expensive to replace the conventional pulping methods.

Ammonia fiber/freeze explosion (AFEX): AFEX is conceptually similar to the steam explosion. In AFEX, the biomass is exposed to hot liquid ammonia under high pressure for a period time, and then the pressure is suddenly released. This swift reduction of pressure opens up the structure of lignocellulosic biomass leading to increased enzymatic hydrolysis of biomass. AFEX pretreatment simultaneously delignify and solubilize some hemicellulose while decrystallizing cellulose, but does not significantly remove hemicellulose as acid and acid-catalyzed steam-explosion pretreatments (Mes-Hartree et al., 1988; Mosier et al., 2005a; Vlasenko et al., 1997). Thus, both micro- and macro-accessibility of cellulose to the cellulase are enhanced. Typical operation conditions of AFEX include aqueous ammonia dosage, (1-2 kg ammonia/kg dry biomass), temperature (<90°C), pH value (<12.0), and pretreatment time (<30 min). The composition of the AFEX pretreated materials was essentially the same as the original (Sun and Cheng, 2002). AFEX has been applied to extensive lignocellulosic materials, including rice straw, municipal solid waste, newspaper, sugar beet pulp, sugar cane bagasse, corn stover, switchgrass, Miscanthus, aspen chips, etc. Herbaceous crops and agricultural residues are well suited for AFEX. However, the AFEX works only moderately and is not attractive for the biomass with high lignin content such as hardwood, softwood and newspaper (McMillan, 1994). The yields of enzymatic hydrolysis of AFEX-pretreated newspaper (18~30% lignin) and aspen chips (25% lignin) were reported as only 40% and below 50%, respectively (McMillan, 1994). The important advantages of AFEX include: (1) producing negligible inhibitors for the downstream biological processes, so water washing process is not necessary (Dale et al., 1984; Mes-Hartree et al., 1988); (2) requiring no particle size reduction (Holtzapple et al., 1990); (3) minimizing formation of sugar degradation products while giving high yields because of the utilization of moderate temperature and pH value during AFEX.

Ammonia must be recycled after the AFEX pretreatment based on the considerations of both the ammonia cost and environmental protection. Therefore, both ammonia cost and the cost of recovery processes drive the cost of this pretreatment (Holtzapple et al., 1992a). An ammonia recovery process was patented by Holtzapple et al. (1992b) who used a superheated ammonia vapor with a temperature up to 200°C to vaporize and strip the residual ammonia in the pretreated biomass and withdraw the evaporated ammonia from the system by a pressure controller for recovery.

pH-controlled liquid hot water: As discussed above, the pH of contents in the reactor can drop to below 4 in the process of LHW pretreatment leading to the inhibitor formation due to the carbohydrate degradation. The pH control for preventing the pH of mixture from falling below 4 limits and controls the secondary chemical reactions (e.g. sugar degradations) occurring during pretreatment (Kohlman et al., 1995). In this pretreatment method, the alkali reagents differ in function from chemicals added as catalysts in alkaline pretreatment methods with its role to maintain the pH constant above 5 and below 7 in order to minimize hydrolysis to monosaccharides (Weil et al., 1998b). Mosier et al. (2005b) pretreated corn fiber using pH-controlled liquid hot water at 160°C and pH above 4.0 and found that 50% of the fiber was dissolved in 20 min. The pretreatment also enabled the subsequent complete enzymatic hydrolysis of the remaining polysaccharides to monosaccharides. The carbohydrates dissolved by the pretreatment were 80% soluble oligosaccharides and 20% monosaccharides with < 1% of the carbohydrates lost to degradation products. Mosier et al (2005c) continued to optimize the pH-controlled liquid hot water for the pretreatment of corn stover and they got the optimal pretreatment conditions to be 190°C for 15 min. At the optimal conditions, 90% of the cellulose was hydrolyzed to glucose by 15FPU cellulase/g-glucan. Both the xylose and the glucose in this undiluted hydrolyzate were shown to be fermented by recombinant yeast 424A (LNH-ST) to ethanol at 88% of theoretical yield. Based on this review, pH-controlled liquid hot water pretreatment technique was just briefly studied for enhancing the enzyme digestibility of cellulosic biomass. It seems that this technology is more promising than the liquid hot water pretreatment, but more research is needed to obtain its performance data and economic feasibility analysis.

Wet oxidation (WO): In WO process, the lignocellulosic materials are treated in water with the presence of oxygen or air at elevated temperature and pressure (McGinnis et al., 1983). WO technique has been used commercially for the production of vanillin from pulped lignin (Salvesen et al., 1952) and for increasing the bulk of municipal solid waste (Guccione, 1964). In addition to applications in the pulping recovery system and wastewater treatment, WO process has been successfully applied to the pretreatment of lignocellulosic biomass to improve biomass digestibility. The primary mechanisms of WO are to solubilize hemicellulose and partial lignin rendering improved enzymatic hydrolysis of residual cellulose. The critical factors of WO include temperature (120-200°C), pressure (0.3-3 MPa), residence time (5-30 min), and catalyst (alkali or acid, pH=3-10). Depending on the operation temperature, two distinct types of reactions occur during WO process: a low-temperature hydrolytic reaction and a high-temperature oxidative reaction (McGinnis et al., 1983). The major reaction at low temperature is initiated by the production of small amount of organic acids. The acids cause the hydrolysis of hemicellulose and partial hydrolysis of lignin leaving behind a solid residue with a high percentage of cellulose. At high temperature, the oxidative reaction makes more hemicellulose and lignin solubilized, and cellulose is partially hydrolyzed. Meanwhile, the formation rate of inhibitors such as carboxylic acids, phenols and furaldehydes is also increased. High oxygen concentration is more effective in increasing both lignin removal and acid formation rate which leads to the enhanced hemicellulose solubilization. The common catalysts include NaOH, Na_2CO_3 and H_2SO_4. The base was believed better than acid as a catalyst in WO process because it can achieve equal even better pretreatment effects than acid with lower level of inhibitors (Martin et al., 2007b).

The major advantages of the WO technique are that: (1) it uses cheap chemicals-water, oxygen (or air), and/or a little alkali or acid; (2) it has little environmental impact; and (3) it produces relatively low inhibitors due to its moderate pretreatment severity. The WO method has been extensively used to pretreat various lignocellulosic materials, including hardwood (McGinnis et al., 1983), softwood (Palonen et al., 2004), sugarcane bagasse (Martin and Thomsen, 2007; Martin et al., 2007b), wheat straw (Bjerre et al., 1996; Felby et al., 2003; Klinke et al., 2002; Schmidt and Thomsen, 1998), corn stover (Varga et al., 2003), rice hull/cassava stalks/peanut shells (Martin and Thomsen, 2007), clover and ryegrass mixtures (Martin et al., 2008), and miscanthus (Sorensen et al., 2008). It was found that the pretreatment effect of WO process varies with lignocellulosic materials (Martin and Thomsen, 2007). It appears that the WO pretreatment on softwood is not as attractive as on hardwood, herbaceous crops or agricultural residue. In addition, the pretreatment effect should be weighed against the formation of inhibitors when the efforts are made on the optimization of pretreatment severity.

Ionic liquids pretreatment (ILs): ILs technique has recently received extensive research attention on the cellulose dissolution (Fukaya et al., 2008; Kosan et al., 2008; Mikkola et al., 2007; Rogers, 2002; Rogers et al., 2004; Swatloski et al., 2003). ILs show promise as efficient and "green" novel cellulose solvents. They can dissolve large amounts of cellulose at considerable mild conditions and feasibility of recovering nearly 100% of the used ILs to their initial purity makes them attractive (Heinze et al., 2005). This technology was used to direct dissolution of cellulose in the commercially Lyocell process as a modern industrial fiber-making, but not for pretreatment of lignocellulosic biomass for ethanol production (Fink et al., 2001). As cellulose solvents, ILs possess several advantages over regular volatile organic solvents, including biodegradability, low toxicity, broad selection of anion and cation combinations, low hydrophobicity, low viscosity, enhanced electrochemical stability, thermal stability, high reaction rates which may lead to great amount of output, low volatility with potentially minimal environmental impact, and non-flammable property (Feng and Chen, 2008). The dissolution mechanism of cellulose in ILs involves the oxygen and hydrogen atoms of cellulose hydroxyl groups in the formation of electron donor-electron acceptor (EDA) complexes which interact with the ILs (Figure 4) (Feng and Chen, 2008). The cations and anions in ILs act as electron acceptor center and electron-donor center, respectively. The two centers must be located close enough in space to permit the interactions and to permit the EDA complexes to form. Upon interaction of the cellulose-OH and ILs, the hydrogen bonds are broken, resulting in opening of the hydrogen bonds between molecular chains of the cellulose (Feng and Chen, 2008). The interaction finally results in the dissolution of cellulose. Solubilized cellulose can be recovered by rapid precipitation with some anti-solvents such as water, ethanol, methanol, or acetone. The recovery cellulose was found to have the same DP and polydispersity as the initial cellulose, but significantly different macro- and micro-structure, especially the decreased degree of crystallinity (Zhu, 2008). The previously used ILs include N-methylmorpholine-N-oxide monohydrate (NMMO) (Kuo and Lee, 2009), 1-n-butyl-3-methylimidazolium chloride (BMIMCl) (Dadi et al., 2006, 2007; Liu and Chen, 2006; Swatloski et al., 2002), 1-allyl-3-methylimidazolium chloride (AMIMCl) (Dadi et al., 2007; Wu et al., 2004; Zhang et al., 2005), 3-methyl-N-bytylpyridinium chloride (MBPCl) and benzyldimethyl (tetradecyl) ammonium chloride (BDTACl) (Heinze et al., 2005).

Although numerous studies have been done on the utilization of ILs on the dissolution of cellulose, few of them are related to the pretreatment of lignocellulosic materials for enhancing the enzymatic digestibility of lignocellulosic materials. NMMO, BMIMCl, and AMIMCl were reported for lignocellulosic biomass pretreatment. Most pretreatment studies were focused on the pure cellulose with high crystallinity such as Avicel. Dadi et al. (2006) found that the initial enzymatic hydrolysis rate and yield of pretreated Avicel PH-101 were increased by 50- and 2-fold, respectively, in comparison with untreated Avicel. Kuo and Lee (2009) also observed that the NMMO pretreated sugarcane bagasse has 2-fold higher enzymatic hydrolysis yield as compared to untreated bagasse. Using BMIMCl to pretreat raw and steam-exploded wheat straw, Liu and Chen (2006) obtained significant improvement of enzymatic hydrolysis yield of wheat straw. They found BMIMCl modified the structure of wheat straw by decreasing the DP and crystallinity and partially solubilizing cellulose and hemicellulose.

Application of ionic liquids has opened new ways for the efficient utilization of lignocellulosic materials in such areas as biomass pretreatment and fractionation. However, there are still many challenges in putting these potential applications into practical use. Such challenges include the high cost of ILs, regeneration requirement, lack of toxicological data and knowledge about basic physico-chemical characteristics, action mode on hemicellulose and/or lignin contents of lignocellulosic materials, and inhibitor generation issues. Further research and financial support are required to address such challenges (Zhu, 2008).

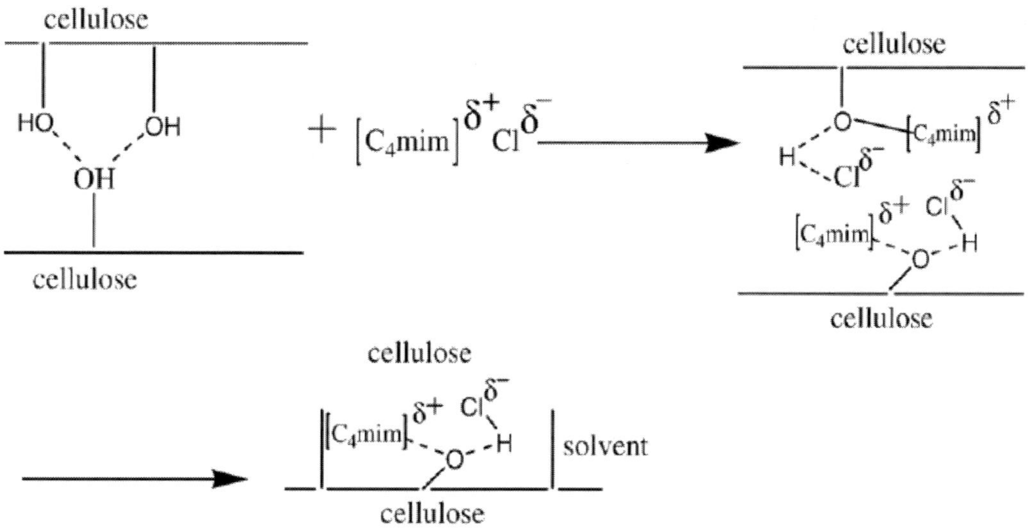

Figure 4. Dissolution mechanism of cellulose in ionic liquids (Adapted from Feng and Chen, 2008).

Other Pretreatments: Many other pretreatment methods had been developed and applied to the pulping and/or pretreatment of lignocellulosic biomass. They include oxidation delignification using oxidative agents (chlorine dioxide, nitrogen oxide, sulfur dioxide, ozone, sodium chlorite, hydrogen peroxide, etc.), cellulose extraction using some solvents (cadoxen, CMCS, concentrated phosphoric acid, triethylamine oxide, etc.), and biomass swelling using some swelling agents (zinc chloride, ruthenium red, sodium zincate, etc.). Most of them are not used any more because of their native drawbacks such as low efficiency and high cost. Refer to Fan et al. (1982) for the information about other pretreatment technologies.

Biological Pretreatment

Biological pretreatment employs wood degrading microorganisms, primarily including white-, brown-, soft-rot fungi, and bacteria to modify the chemical composition and/or structure of the lignocellulosic biomass so that the biomass is more amenable to enzyme digestion (Ander and Eriksson, 1978; Hatakka, 1984; Kurakake et al., 2007; Lee et al., 2007; Singh, et al., 2008). According to the previous research, fungi have distinct degradation characteristics on lignocellulosic biomass. In general, brown and soft rots mainly attack cellulose while imparting minor modifications to lignin; however, white-rot fungi are more actively degrade the lignin component (Schurz 1978). White-rot fungi were considered the most promising basidiomycetes for lignocellulosic bio-pretreatment (Ander and Eriksson, 1978). Hwang et al. (2008) studied biological pretreatment of wood chips using four different white-rot fungi for 30 days and found that the glucose yield of pretreated wood by *Trametes versicolor* MrP 1 reached 45% by enzymatic hydrolysis while 35% solid was converted to glucose during fungi incubation. They suggested that the selection of proper wood species and fungal strains and optimization of glucose recovery are all necessary for the fungal pretreatment of woody biomass, which is used as a carbon source by fungi. A Japanese red pine *Pinus densiflora* (softwood) was pretreated biologically by white-rot fungi of *Ceriporia lacerata*, *Stereum hirsutum*, and *Polyporus brumalis* and it was found that *S. hirsutum* was the most effective to degrade lignin and improved the enzymatic digestibility of wood (Lee et al., 2007). After eight weeks of pretreatment with *S. hirsutum*, total weight, lignin, and holocellulose losses were 10.7%, 14.52%, and 7.81%, respectively. Studying the enzyme system of fungi, they also observed that total weight loss and changes in chemical composition of biomass were well correlated with the enzyme activities which are linked to lignin- and cellulose degradation in these fungi. Keller et al. (2003) reported that their preliminary tests showed a 3- to 5-fold improvement in enzymatic digestibility of corn stover after pretreatment with *Cyathus stercoreus*; and a 10- to 100-fold reduction in shear force needed to obtain the same shear rate of 3.2 to 7 rev/s, respectively, after pretreatment with *Phanerochaete chrysosporium*. Zhang et al. (2007b, c) screened 35 isolates of white-rot fungi for the biological pretreatment of bamboo for enzymatic saccharification of bamboo. They observed that *Echinodontium taxodii* 2538, *Trametes versicolor* G20, and *Coriolus versicolor* B1 are the most promising white-rot fungi for highly selective lignin degradation and significant improvement of enzymatic saccharification. The lignolytic white-rot fungi digest lignin by producing lignin-degrading enzymes, which might contain lignin peroxidases, manganese-dependent peroxidases, polyphenol oxidases, laccases, H_2O_2 producing enzymes, or quinone-reducing enzymes (Blanchette, 1991; Boominathan and Reddy, 1992; Kirk and Farrell, 1987; Waldner et al., 1988). Degradation of lignin by white-rot fungi is a co-oxidative process, thus accompanying carbon source is necessary, usually from cellulose and/or hemicellulose. In order to prevent the carbohydrate loss such as cellulose, the fungal strains with preference of lignin degradation such as *P. ostreatus*, *C. subvermispora*, *Cyathus stercoreus* or genetically modified *P. chrysosporium* to produce less cellulase activity (Anonymous, 1980; Kerem et al., 1992; Kirk et al., 1986). Eriksson and Goodell (1974) and Eriksson (1978) exposed *Potyporus adustus* to UV light to destroy a regulatory gene coding the synthesis of cellulase, mannanase, and xylanase so that they achieved almost specific lignin degradation cellulase-less mutants of white-rot fungi. Ander and Eriksson (1977, 1978) also developed a cellulase-less mutant of *Sporotrichum pulverulentum* to reduce the lignin content of birch rods by 17 % and stiffness by 10 % after incubated the fungi on the birch rods for six weeks. Taniguchi et al. (2005) found that the use of these alternative strains generally requires longer pretreatment time.

The biological pretreatment appears to be a promising technique and has very evident advantages, including no chemical requirement, low energy input, mild environmental conditions, and environmentally friendly working manner (Hsu, 1996; Sun and Cheng, 2002). However, its disadvantages are as evident as its advantages since biological pretreatment is very slow (this is a major drawback) and requires careful control of growth conditions and large amount of space to perform treatment (Chandra et al., 2007; Fan et al., 1982; Hsu, 1996). In addition, most lignolytic microorganisms solubilize/consume not only lignin but also hemicellulose and cellulose (Basaglia et al., 1992; Eriksson et al., 1980; Kohlmann et al., 1993; Kuhar et al., 2008; Lee et al., 2007; Shi et al., 2008; Singh et al., 2008). Therefore, the biological pretreatment faces technoeconomic challenges and is less attractive commercially. Thus far, the studies of biological pretreatment have been limited in bench-scale. Although, a fair amount of work has been done in this field, the primary application of biological pretreatment has been employed as a biopulping option for the pulp and paper industry rather than as a pretreatment for biofuel applications (Chandra et al., 2007). However, it is possible to accelerate its rate through genetic modification of lignin degrading microorganisms or through the combination of partial physical and/or chemical pre-pretreatment with biological pretreatment (Fan et al., 1982).

Combination Pretreatment

Combination pretreatment technique usually incorporates two or more steps consisting over one single pretreatment method from the same or different categories. The combination profiles could be two physical pretreatments conducted in sequence (Li et al., 2007; Pritchard et al., 1962; Yang et al., 2008), combination of physical pretreatment and chemical pretreatment (Chosdu et al., 1993; Corredor et al., 2008; Li et al., 2004, 2007; Lu and Kumakura, 1995; Sorensen et al., 2008; vanWyk, 1997; Xin and Kumakura, 1993; Yoon, 1998; Zhu et al., 2005c), two chemical pretreatments applied in sequence (Bjerre, et al., 1996; Chammem et al., 2005; Kim et al., 2001; Pan et al., 2006; Soderstrom et al., 2002, 2003a, b; Teixeira et al., 1999a, b, 2000), a chemical pretreatment followed by a biological pretreatment (Basaglia et al., 1992; Yu et al., 2009), and a physical pretreatment followed by a biological pretreatment (Kohlmann et al., 1995; Yu et al., 2009). The general goal of combination pretreatment is to achieve better pretreatment results than a single method so that the involved individual pretreatment units should be expected to work synergistically. However, Schwald et al. (1989a, b) found a second pretreatment step did not improve the digestibility of substrates pretreated by a first step. Among these combination pretreatment methods, the combinations of physical pretreatment and chemical pretreatment are the most effective combination, and have received the most attention in recent years. Compared to the pretreatment method with a single step, the capital and processing costs of a combination pretreatment method might usually involve additional equipment and/or additional chemicals. The benefits of the combination pretreatment must be balanced against the extra costs.

Characteristics of Pretreatment Product Streams

Different pretreatment technique has different action mode so that different product streams will be produced and will probably require different down stream hydrolysis and fermentation strategies. Table 3 shows the various product streams derived from three major lignocellulosic biomass components (cellulose, hemicellulose and lignin) as a result of pretreatment.

The product streams shown in Table 3 are based on a single pretreatment method. If combination pretreatment methods are employed to pretreat biomass, corresponding down stream process strategies might be different, e.g., fractionation of three components of lignocellulosic materials into three separated product streams with hemicellulose removal, delignification and cellulose treatment in that order. Different combination scheme might result in different product streams, but the process goal and economic feasible need to be critically considered. In addition, the formation of inhibitory compounds is a critical issue for the pretreatment step. Some acids, phenolics, and carbohydrate/lignin degradation products were found or suspected to inhibit the enzymatic hydrolysis, fermentation or both. Therefore, the detoxification techniques might be devised to reduce or eliminate the inhibition if the level of inhibitor is unavoidable.

Table 3. Product streams of different pretreatment techniques

Product stream	Pretreatment technique	Fermentation strategies
Cellulose, hemicellulose, and lignin in one product stream	Mechanical comminution, irradiation, biological pretreatments, and ionic liquids	Co-fermentation or sequential fermentation of pentose and hexose if no further pretreatments are used.
Solubilized lignin and hemicellulose sugars in liquid phase and cellulose in solid phase	Alkali, organosolv, AFEX, ARP, wet oxidation	With liquid and solid separation, fermentation of pentose or co-fermentation of hexose and pentose depending on the substrate. The solid stream will go through hydrolysis and sequential or co-fermentation.
		Without liquid and solid separation, solubilized oligomeric hemicellulose sugars in the liquid might need depolymerization. The solid cellulose portion will need hydrolysis. Then, the pentose and hexose will undergo co-fermentation or sequential fermentation.
Solubilized hemicellulose sugars in liquid phase and lignin and cellulose in solid phase	Uncatalyzed and catalyzed (SO_2, CO_2) steam-explosion, liquid hot water, pH-controlled liquid hot water, dilute acid pretreatment, wet oxidation	With liquid and solid separation, liquid might need pentose or both pentose and hexose fermentation. The solid cellulose will be hydrolyzed and fermented into ethanol straightforward.
		Without liquid and solid separation, solid hydrolysis is need first. Depending on the pretreatment severity, the solubilized hemicellulose oligomeric sugars might need depolymerization first. And then mixture of pentose and hexose need sequential or co-fermentation.

Current Status and Challenges

Biomass pretreatment, for the purpose of improved enzymatic digestibility and/or fermentability, has been studied for years. Although extensive research has been done to develop numerous effective pretreatment techniques on various lignocellulosic biomass feedstocks, none has been really commercialized for cellulosic ethanol production. Only a few of the reported pretreatment methods such as dilute acid and steam-explosion have been tested on a demonstration scale. One of the most famous demonstration scale cellulose ethanol facilities is held by Iogen Corporation in Canada. Iogen Corporation is using modified steam-explosion pretreatment to improve the enzymatic digestibility of straw to produce cellulosic ethanol with yield of 50 M ton/year. DOE has also been supporting several biofuel programs through bioenergy laboratories or centers such as NREL, GLBRC, and OREL to develop various pretreatment methods, collect process performance date, and evaluate their economic feasibility. Currently, the research of pretreatment is facing several technoeconomic challenges, including environmental impact, delignification, co-fermentation of hexose and pentose into ethanol, hydrolyzate detoxification, hemicellulose and cellulose separation, pretreatment product digestibility, energy demand, and processing costs. Future research should address these challenges to promote the commercialization of cellulosic ethanol.

FUTURE RESEARCH

Extensive research has been done on the development of advanced pretreatment technologies on various lignocellulosic materials for cellulosic ethanol production. The general goal of these efforts is to prepare more digestible biomass to ease bioconversion of biomass into cellulosic ethanol. An ideal cost-effective pretreatment method might have several characteristics: (1) maximum fermentable carbohydrate recovery; (2) minimum inhibitors produced from carbohydrate degradation during pretreatment; (3) low environmental impact; (4) low demand of post-pretreatment processes such as washing, neutralization, and detoxification; (5) minimum water and chemical use; (6) low capitol cost of equipments; (7) moderately low energy input; (8) relatively fast treatment rate; and (9) production of high value-added by-products. Therefore, the future research would be focused on the following areas. First, reduction of water and chemical use: the increased concern of environmental protection, water saving, human heath, and waste treatment, effective pretreatment that use few or no chemicals/water and produce few or no pollutants are likely to receive research, development, and commercial momentum; Second, recovery of carbohydrates and by-products. Both cellulose and hemicellulose sugars could be converted into ethanol. Lignin, a kind of by-product could be used as boiler fuel or the source for special chemicals. Hemicellulose and by-product credits will likely be needed to render the cellulosic ethanol production process more economically feasible. Therefore, pretreatment techniques that can efficiently fractionate biomass into multiple streams containing fermentable sugars and value-added by-products in concentrations that make purification, utilization and/or recovery economically practical are likely to gain research interest; Third, development of clean delignification. Many researchers have been studying microorganisms that can co-ferment 5C and 6C sugars into ethanol. Lynd et al. (2005) believed the Consolidated Bioprocessing (CBP) is the logic end-point of biomass processing which involves such co-fermenting microorganisms. Therefore, future research would be driven by the development of selective lignin-removal pretreatment techniques, remaining cellulose and hemicellulose intact. The delignification process developed by Avgerinos and Wang (1983) is an example. Benefits should be weighed against the economics of such processes; Fourth, fundamental understanding of pretreatment mechanisms and the relationship between the biomass structure features and enzymatic hydrolysis. The pretreatment process is usually time-consuming and labor intensive. The datum collecting and analysis is also a big challenge. A predictive pretreatment model will be a very useful forecasting tool under this context. It will enable the selection, design, optimization, process control, and economic analysis of a pretreatment technology that match specific lignocellulosic biomass feestocks with the suitable method and process pattern; Fifth, reduction of the generation of inhibitors. Sever common pretreatment by-products such as furfural, 5-HMF and acetic acid are proved to be strong inhibitors for fermentation and/or enzymatic hydrolysis. The costly detoxification processes are usually necessary to prevent the inhibition. As a result, the development of pretreatment methods with low inhibitor production will be likely a research direction.

REFERENCES

Abatzoglou, N.; Chornet, E.; Belkacemi, I.; Overend, R. *Chem. Eng. Sci.* 1992, 47, 1109-1122.

Abdi, N.; Hamadache, F.; Belhocine, D.; Grib, H.; Lounici, H.; Piron, D. L.; Mameri, N. *Biochem. Eng. J.* 2000, 6, 177-183.

Adler, E. *Wood Sci. Technol.* 1977, 11, 169-218.

Allen, S. G.; Kam, L. C.; Zemann, A. J.; Antal Jr. M. J. Presented at 207 the ACS National Meeting, San Diego, CA, 1994, March, 13-17.

Allen, S. G.; Kam, L. C.; Zemann, A. J.; Antal Jr., M. *J. Ind. Eng. Chem. Res.* 1996, 35, 2709-2715.

Allen, S. G.; Schulman, D.; Lichwa, J.; Antal, M. *J. Ind. Eng. Chem. Res.* 2001, 40, 2934-2941.

Ander, P.; Eriksson, K. -E. *Physiol. Plant.* 1977, 41, 239-248.

Ander, P.; Eriksson, K. E. In: *Progress in Ind. Microbiol.*; Bull, M. J.; Ed.; Elsevier: New York, NY, 1978; Vol. 14, pp 1.

Anonymous. Gasohol U.S.A. 1980, 2, 26.

Antal Jr., M. J. In *Proceedings of the 12nd International Conference on the Properties of Water and Steam;* White, Jr., H. J.; Ed.; Begell House: New York, NY, 1996; pp 24-32.

Araque, E.; Parra, C.; Freer, J.; Contreras, D.; Rodriguez, J.; Mendonca, R.; Baeza, J. Enzyme Microb. Technol. 2008, 43, 214-219.

Aristidou, A.; Penttila, M. *Curr. Opin. Biotechnol.* 2000, 11, 187-198.

Aronovsky, S. I.; Gortner, R.A. *Ind. Eng. Chem.* 1930, 22, 264-274.

Avellar, B. K.; Glasser, W.G. *Biomass Bioenergy* 1998, 14, 205-218.

Avgerinos, G. C.; Wang, D. I. C. *Biotechnol. Bioeng.* 1983, 25, 67-83.

Aziz, S.; Sarkanen, K. Tappi. J. 1989; 72, 169-175.

Azzam, A. M. J. *Environ. Sci. Health.*, Part B 1989, 24, 421-433.

Ballesteros, I.; Oliva, J. M.; Negro, M. J.; Manzanares, P.; Ballesteros, M. *Process Biochem.* 2002, 38, 187-192.

Ballesteros, I.; Ballesteros, M.; Manzanares, P.; Negro, M. J.; Oliva, J. M.; Saez, F. *Biochem. Eng. J.* 2008, 42, 84-91.

Bak, J. S.; Ko, J. K.; Han, Y. H.; Lee, B. C.; Choi, I. G.; Kim, K. H. *Bioresour. Technol.* 2009, 100, 1285-1290.

Basaglia, M.; Concheri, G.; Cardinali, S.; Pasti-Grigsby, M. B.; Nuti, M. P. *Can. J. Microbiol.* 1992, 38, 1022-1025.

Baugh, K. D.; Levy, J. A.; McCarty, P. L. *Biotechnol. Bioeng.* 1988a, 31, 50-61.

Baugh, K. D.; Levy, J.A.; McCarty, P. L. *Biotechnol. Bioeng.* 1988b, 31, 62-70.

Beardmore, D. H.; Fan, L. T.; Lee, Y. H. *Biotechnol. Lett.* 1980, 2, 435-438.

Beery, K.; Hendrickson, R.; Brewer, M.; Mosier, N.; Dien, B.; Dreschel, R.; Welch, G.; Bothast, R.; Ladisch, M. BTEC Paper 46, 219th National Meeting of the ACS, San Francisco, CA; Biotechnology Secretariat (BTEC), *Biobased Processing to Chemicals*; 2000,V: Process Engineering.

Bennani, A.; Rigal, L.; Gaset, A. *Biomass Bioenergy* 1991, 1, 289-296.

Bertran, M. S.; Dale, B. F. *Biotechnol. Bioeng.* 1985, 27, 177-181.

Betrabet, S. M.; Paralikar, K. M. *Cellul. Chem. Technol.* 1977, 11, 615-625.

Biermann, C.J.; Schultz, T.P.; McGinnis, G. D. *J. Wood Chem. Technol.* 1984, 4, 111-128.

BiofuelsDigest. (2008). http://biofuelsdigest.com/blog2/2008/01/09/us-ethanol-production-capacity-to-increase-to-133-billion-gallons-in-2008/

Bjerre, A. B.; Olesen, A. B.; Fernqvist, T. *Biotechnol. Bioeng.* 1996, 49, 568-577.

Blanchette, R. A. *Annu. Rev. Phytopathol.* 1991, 29, 381-398.

Blouin, F. A.; Arthur, J. C. *Text. Res. J.* 1958, 28, 198-204.

Blouin, F. A.; Arthur, J. C. *J. Chem. Eng. Data* 1960, 5, 470.

Bobleter, O.; Niesner, R.; Rohr, M. *Appl. Polym. Sci.* 1976, 20, 2083-2093.

Bobleter, O.; Concin, R. *Cellul. Chem. Technol.* 1979, 13, 583-593.

Bobleter, O.; Binder, H.; Concin, R.; Burtscher, E. In *Energy from Biomass*; Palz, W.; Chartier, P.; Hall, D.O.; Eds.; *Applied Science Publishers*: London, 1981; pp 554-562.

Bobleter, O. *Prog. Polym. Sci.* 1994, 19, 797-841.

Bonn, G.; Concin, R.; Bobleter O. *Wood Sci. Technol.* 1983, 17, 195-202.

Boominathan, K.; Reddy, C. A. *Proc. Natl. Acad. Sci.* (USA) 1992, 89, 5586-5590.

Bothast, R. J.; Nichols, N. N.; Dien, B. S.; Cotta, M. A. Biocatalysts for production of bioethanol. *Society of Industrial Microbiology Annual Meeting*, 2002, paper No. 12-1.

Bouchard, J.; Nguyen, T. S.; Chornet, E.; Overend, R. P. *Biomass* (UK), 1990, 23, 243-261.

Bouchard, J.; Nguyen, T. S.; Chornet, E.; Overend, R. P. *Bioresour. Technol.* 1991, 36, 121-131.

Boussaid, A.; Robinson, J.; Cai, Y. J.; Gregg, D. J.; Saddler, J. R. *Biotechnol. Bioeng.* 1999, 64, 284-289.

Boussaid A. L.; Esteghlalian, A. R.; Gregg, D. J.; Lee, K. H.; Saddler, J. N. *Appl. Biochem. Biotechnol.* 2000, 84/86, 693.

Bradbury, A. G. W.; Sakai, Y.; Shafizadeh, F. *J. Appl. Polym. Sci.* 1979, 23, 3271-3280.

Brandon, S. K.; Eiteman, M. A.; Patel, K.; Richbourg, M. M.; Miller, D. J.; Anderson, W. F.; Peterson, J. D. *J. Chem. Technol. Biotechnol.* 2008, 83, 505-512.

Brennan, A. H.; Hoagland, W.; Schell, D. *J. Biotechnol. Bioeng. Symp.* 1986, 17, 53-70.

Brenner, W. In New Approaches for the Acid Hydrolysis of Cellulose, *Dept. of Appl. Sci.*, New York Univ., NYU/DAS-77-30, New York, 1977.

Brink, D. L. Method of treating biomass material. 1993. *U.S. Patent* 5,221,357.

Brink, D. L. Method of treating biomass material. 1994. *U.S. Patent* 5,366,558.

Brownell, H. H.; Saddler, J. N. *Biotechnol. Bioeng. Symp.* 1984, 14, 55-68.

Brownell, H. H.; Yu, E. K. C.; Saddler, J. N. *Biotechnol. Bioeng*. 1986, 28, 792-801.

Browning, B.L. *Methods of wood chemistry*; Interscience: New York, 1967.

Bura, R. Bioconversion of corn fiber to ethanol. Ph.D. Thesis, University of British Columbia, Vancouver, Canada; 2004.

Cadoche, L.; Lopez, G. D. *Biol. Waste*. 1989, 30, 153-157.

Cahela, D. R.; Lee, Y. Y.; Chambers, R. P. *Biotechnol. Bioeng*. 1983, 25, 3-17.

Cantarella, M.; Cantarella, L.; Gallifuoco, A.; Spera, A.; Alfani, F. *Biotechnol. Progr*. 2004, 20, 200-206.

Capekmenard, E.; Jollez, P.; Chornet, E.; Wayman, M.; Doan, K. *Biotechnol. Lett*. 1992, 14, 985-988.

Carrasco, F.; Roy, C. *Wood Sci. Technol*. 1992, 26, 189-208.

Carrasco, J. E.; Saiz, M. C.; Navarro, A.; Soriano, P.; Saez, F.; Martinez, J. M. *Appl. Biochem. Biotechnol*. 1994, 45-46, 23-34.

Carrillo F.; Lis, M. J.; Colom, X.; Lopez-Mesas, M.; Valldeperas, *J. Process Biochem*. 2005, 40, 3360-3364.

Caulfield, D. F.; Moore, W. E. *Wood Sci*. 1974, 6, 375-379.

Chammem, N.; Kachouri, M.; Mejri, M.; Peres, C.; Boudabous, A.; Hamdi, M. *Bioresour Technol*. 2005, 96, 1311-1316.

Chandra, R. P.; Bura, R.; Mabee, W. E.; Berlin, A.; Pan, X.; Saddler, J. N. *Adv. Biochem. Engin/Biotechnol*. 2007, 108, 67-93.

Chang, M. M.; Chou, T. Y. C.; Tsao, G. T. *Adv. Biochem. Eng*. 1981, 20, 15-42.

Chang, V. S.; Burr, B.; Holtzapple, M. T. *Appl. Biochem. Biotechnol*. 1997, 63/65, 3-19.

Chang, V. S.; Nagwani, M.; Holtzapple, M. T. *Appl. Biochem. Biotechnol*. 1998, 74, 135-159.

Chang, V., Holtzapple, M. Fundamental factors affecting biomass enzymatic reactivity. *Appl. Biochem. Biotechnol*. 2000, 84/86, 5–37.

Chang, V. S.; Nagwani, M.; Kim, C. H.; Holtzapple, M. T. *Appl. Biochem. Biotechnol*. 2001, 94, 1-28.

Chosdu, R.; Hilmy, N.; Erizal-Erlinda, T. B.; Abbas, B. Radiation and Chemical Pretreatment of Cellulosic Waste. *Radiat. Phys. Chem*. 1993, 42, 695-698.

Chum, H.L.; Douglas, L.J.; Feinberg, D.A.; Schroeder, H.A. Evaluation of pretreatments of biomass for enzymatic hydrolysis of cellulose. *Solar Energy Research Institute*: Golden, CO, 1985; pp. 1-64.

Chum, H. L.; Black, S.; Johnson, D. K.; Sarkanen, K.V. In *Proceedings of the Solvent Pulping-Promises and Problems Conference*; Doshi and Associates, Inc.: Appleton, Wisconsin, 1987, pp. 35-41.

Chum, H. L.; Johnsoon, D. K.; Black, S. *Biotechnol. Bioeng*. 1988, 31, 643-649.

Chum, H. L.; Johnson, D. K.; Black, S. K. *Ind. Eng. Chem. Res*. 1990, 29, 156-162.

Chundawat, S. P. S.; Venkatesh, B.; Dale, B. E. *Biotechnol. Bioeng*. 2007, 96, 219-231.

Clark, T.; Mackie, K. L. *J. Chem. Biotechnol*. 1984, 34B, 101.

Clark, T. A.; Mackie, K. L. *J. Wood Chem. Technol*. 1987, 7, 373-403.

Clarke, S. D.; Dyer, I. A. *J. Anim. Sci*. 1973, 37, 1022.

Converse, A. O.; Kwarteng, I. K.; Grethlein, H. E.; Ooshima, H. *Appl. Biochem. Biotechnol*. 1989, 20/21, 63-78.

Converse, A.O. In *Bioconversion of forest and agricultural plant residues*; Saddler, J. N.; Ed.; CAB International: Wallingford, 1993, pp 93-106.

Corredor, D. Y.; Sun, X. S.; Salazar, J. M.; Hohn, K. L.; Wang, D. *J. Biobased Mater. Bioenergy* 2008, 2, 43-50.

Cowling, F. B.; Brown, W. In *Celhdases and their Applications*, Bould, R. F.; Ed.; ACS: Washington DC, 1969, pp 152-187.

Cowling, E. B.; Kirk, T. K. *Biotechnol. Bioeng. Symp.* 1976, 6, 95-123.

Dadi, A. P.; Schall, C. A.; Varanasi, S. *Biotechnol. Bioeng.* 2006, 95, 904-910.

Dadi, A. P.; Schall, C. A.; Varanasi, S. *Appl. Biochem. Biotechnol.* 2007, 137, 407-421.

Dale, B. E.; Moreira, M. *J. Biotech. Bioeng. Symp.*, 1982, 12, 31-43.

Dale, B. E.; Henk, L. L.; Shiang, M. Dev. *Ind. Microbiol.* 1984, 26, 223-233.

Deanda, K.; Zhang, M.; Eddy, C.; Picataggio, S. *Appl. Environ. Microbial.* 1996, 62, 4465-4470.

De Bari, I.; Nanna, F.; Braccio, G. *Ind. Eng. Chem. Res.* 2007, 46, 7711-7720.

Delgenes, J. P.; Moletta, R.; Navarro, J. M. *Enzyme Microb. Technol.* 1996, 19, 220-225.

DeLong, E.A. 1981. *Canadian Patent* 1,096,374.

Detroy, R. W.; Lindenfelser, L. A.; Julian, G. St.; Orton, W. L. *Biotech. Bioeng. Symp.* 1980, 10, 135-148.

Dien, B. S.; Hespell, R. B.; Wyckoff, H. A.; Bothast, R. *J. Enzyme Microb. Technol.* 1998, 23, 366-371

Dien, B. S.; Cotta, M. A.; Jeffries, T. W. *Appl. Microbiol. Biotechnol.* 2003, 63, 258-266.

Dien, B. S.; Li, X. L.; Iten, L. B.; Jordan, D. B.; O'Bryan, P. J.; Cotta, M. A. *Enzyme Microb. Technol.* 2006, 39, 1137-1144.

Draude, K. M.; Kurniawan, C. B.; Duff, S. T. B. *Bioresour. Technol.* 2001, 79, 113-120.

Duff, S. J. B.; Murray, W. D. *Bioresour. Technol.* 1996, 55, 1-33.

Dunlap, C. E.; Thomson, J.; Chiang, L. C. *AIChE Syposium Series.* 1976, 72, 58.

Dunlap, C. E.; Chiang, L. C. In: Cellulose Degradation A Common Link; Shuler, M. C.; Ed.; 1980, C.R.C. Press.

Eggeman, T.; Elander, R. T. *Bioresour. Technol.* 2005, 96, 2019-2025.

Emsley, A. M.; Stevens, G. C. *Cellulose* 1994, 1, 26-56.

Engstrom, A. C.; Ek, M.; Henriksson, G. *Biomacromolecules*, 2006, 7, 2027-2031.

Eriksson, K. E.; Goodell, E. W. *Can. J. Microbiol.* 1974, 20, 371.

Eriksson, O.; Lindgren, B. *Svensk Papperstidning* 1977, 80, 59-63.

Eriksson, K. E. In Bioconversion of Cellulosic Substrates into Energy Chemicals and Microbial Protein Syrup. Proc., Ghose, T. K.; Ed.; New Delhi: IIT Delhi, 1978, pp 195.

Eriksson, K.; Grunewald., A.; Vallander, L. *Biotech. Bioeng.* 1980, 22, 363-376.

Esteghlalian, A.; Hashimoto, A.G.; Fenske, J. J.; Penner, M.H. *Bioresour. Technol.* 1997, 59, 129-136.

Esteghlalian, A. R.; Bilodeau, M.; Mansfield, S. D.; Saddler, J. N. *Biotechnol Progr* 2001, 17, 1049.

Excoffier, G.; Toussaint, B.; Vignon, M. R. *Biotech. Bioeng.* 1991, 38, 1308-1317.

Fan, L. T.; Lee, Y. –H.; Beardmore, D. H. *Biotechnol. Bioeng.* 1980a, 22, 177-199.

Fan, L. T.; Lee, Y. H.; Beardmore, D. H. *Presented at lind Int. Syrup. on Bioconv. and Biochem.* Eng., IIT, Delhi, India, 1980b, March, 3-6.

Fan, L. T.; Lee, Y-. H.; Beardmore, D. R. *Biotechnol. Bioeng.* 1981a, 23, 419-424.

Fan, L.T.; Gharpuray, M. M.; Lee, Y. *Biotech. Bioeng. Symp.*, 1981b, 11, 29-45.

Fan, L. T.; Lee, Y.; Gharpuray, M. M. Adv. *Biochem. Eng.* 1982, 23, 157-187.

Fan, L.T.; Gharpuray, M. M.; Lee, Y. -H. In *Cellulose Hydrolysis Biotechnology Monographs*. Springer: Berlin, 1987, p 57.

Farid, M. A.; Shaker, H. M.; Eldiwany, A. I. *Enzyme Microb. Technol*. 1983, 5, 421-424.

Fein, J. E.; Potts, D.; Good, D.; Beavan, M.; O'Boyle, A.; Dahlgren, D.; Beck, M. J.; Griffith, R. L. *Energy Biomass and Waste*. 1991, 15, 745-765.

Feist, W. C.; Baker, A. J.; Tarkow, H. *J. Ani. Sci*. 1970, 30, 832.

Felby, C.; Klinke, H. B.; Olsen, H. S.; Thomsen, A. B. In *Applications of Enzymes to Lignocellulosics;* Mansfield, S. D.; Saddler, J. D.; Ed.; ACS Symp. Ser., No. 855; Oxford University Press Canada: Don Mills, ON, 2003; 855, 157-174.

Feng, L.; Chen, Z. *J. Mol. Liq*. 2008, 142, 1-5.

Fernandez-Bolanos, J.; Felizon, B.; Heredia, A.; Rodriguez, R.; Guillen, R.; Jimenez, A. *Bioresour. Technol*. 2001, 79, 53.

Fierobe, H. P.; Bayer, E. A.; Tardif, C.; Czjzek, M.; Mechaly, A.; Belaich, A.; Lamed, R.; Shoham, Y.; Belaich, J. P. *J. Biol. Chem*. 2002, 277, 49621-49630.

Fink, H. P.; Weigel, P.; Purz, H. J.; Ganster, *J. Progr. Polym. Sci*. 2001, 26, 1473-1524.

Foster, B. L.; Dale, B. E.; Doran-Peterson, J. B. *Appl. Biochem. Biotechnol*. 2001, 91/93, 269-282.

Fukaya, Y.; Hayashi, K.; Wada, M.; Ohno, H. *Green Chem*. 2008, 10, 44-46.

Garcia-Aparicio, M. P.; Ballesteros, I.; Gonzalez, A.; Oliva, J. M.; Ballesteros, M.; Negro, M. *J. Appl. Biochem. Biotechnol*. 2006, 129, 278-288.

Gardner, P. T.; Wood, T. J.; Chesson, A.; Stuchbury, T. *J. Sci. Food Agric*. 1999, 79, 18.

Gharpuray, M. M.; Lee, Y. H.; Fan, L. T. *Biotechnol. Bioeng*. 1983a, 25, 157-72.

Gharpuray, M. M.; Fan, L.T.; Lee, Y. In *Wood and Agricultural Residues-Research on Use for Feed, Fuels, and Chemicals; Soltes*, E. J.; Ed.; Academic Press: New York, 1993, pp 369-389.

Ghose, T. K.; Ghosh, P. Bioconversion of cellulosic substances. *J. Appl. Chem. Biotechnol*. 1978, 28, 309-320.

Gilkes, N. R.; Jervis, E.; Henrissat, B.; Tekant, B.; Miller, R. C. Jr.; Warren, R. A. J.; Kilburn, D. G. *J. Biol. Chem*. 1992, 267, 6743-6749.

Glasser, W. G.; Wright, R.S. *Biomass Bioenergy* 1998, 14, 219-235.

Goldstein, I. R.; Pereira, H.; Pittman, L. J.; Strouse, B. A.; Scaringelli, F. P. *Biotech. Bioeng. Symp*. 1983, 13, 17-25.

Gould, J. M. *Biotech. Bioeng*. 1984, 26, 46-52.

Gould, J. M. *Biotech. Bioeng*. 1985, 27, 225-231.

Gracheck, S. J.; Rivers, D. B.; Woodford, L. C.; Giddings, K. E.; Emert, G. H. *Biotech. Bioeng. Symp*. 1981, 11, 47-65.

Grethlein, H. E.; Allen, D. C.; Converse, A.O. *Biotech. Bioeng*., 1984, 26, 1498-1505.

Grethlein, H.E. *Bio/Tech*. 1985, 3, 155-160.

Grethlein, H. E. *Enzyme Microb. Technol*. 1986, 8, 274-280.

Grethlein, H. E.; Converse, A. O. *Bioresour. Technol*. 1991, 36, 77.

Grohmann, K.; Himmel, M.; Rivard, C.; Tucker, M.; Baker, J.; Torget, R.; Graboski, M. *Biotechnol. Bioeng. Symp*. 1984, 14, 137-157.

Grohmann, K.; Torget, R.; Himmel, M. *Biotechnol. Bioeng. Symp*. 1985, 15, 59-80.

Grohmann, K.; Torget, R.; Himmel, M. *Biotechnol. Bioeng. Symp*. 1986, 17, 135-151.

Grohmann, K.; Mitchell, D. J.; Himmel, M. E. *Appl. Biochem. Biotechnol*. 1989, 20/21, 45-61.

Grohmann, K.; Torget, R. W. 1992. *U.S. Patent* 5,125,977.

Grous, W. R.; Converse, A, O.; Grethlein, H, E. *Enzyme Microb. Technol.* 1986, 8, 274-280.

Guccione, E. *Chem. Eng.* 1964, 71, 118.

Gulati, M.; Westgate, P. J.; Brewer, M.; Hendrickson, R.; Ladisch, M. R. *Appl. Biochem. Biotechnol.* 1996, 57/58, 103-119.

Han, Y. W.; Callihan, C. D. *Appl. Microbiol.* 1974, 27, 159-165.

Han, Y. W. *Adv. Appl. Microbial.* 1978, 23, 119-153.

Han, Y. W.; Chen, W. P.; Miles, T. R. *Biotech. Bioeng.* 1978, 20, 567-575.

Han, Y. W.; Ciegler, A. *Annu. Rep. Ferment. Processes* 1983, 6, 299-322.

Harris, E. E.; Kline, A. A. *J. Phys. Colloid Chem.*, 1949, 53, 344-351.

Hatakka, A. I. *Appl. Biochem. Biotechnol.* 1984, 9, 363-364.

Heinze, T.; Schwikal, K.; Barthel, S. *Macromol. Biosci.* 2005, 5, 520-525.

Heitz, M.; Capek-Menard, E.; Koeberle, P. G.; Gagne, J.; Chornet, E.; Overend, R. P.; Taylor, J. D.; Yu, E. *Bioresour. Technol.* 1991, 35, 23-32.

Henrissat, B.; Vigny, B.; Perez, S. *FEBS Lett.* 1988, 231, 177-182.

Hernadi, A. *Celhdose Chem. Technol.* 1984, 18, 115-24.

Herrera, A.; Tellez-Luis, S. J.; Ramirez, J. A.; Vazquez, M. *J. Cereal Sci.* 2003, 37, 267-274.

Hinman, N. D.; Schell, D. J.; Riley, C. J.; Bergeron, P. W.; Walter, P. *J. Appl. Biochem. Biotechnol.* 1992, 34/35, 639-649.

Ho, N. W. Y.; Chen, Z.; Brainard, A. *Appl. Environ. Microbiol.* 1998, 64, 1852-1859.

Ho, N. W. Y.; Chen, Z.; Brainard, A. P.; Sedlak, M. *Adv. Biochem. Eng./Biotechnol.* 1999, 65, 163-192.

Hohberg, A. I.; Aguilera, J. M.; Agosin, A.; San Martin, R. *Biomass*, 1989, 18, 81-93.

Holtzapple, M. T.; Jun, J-H.; Ashok, G.; Patibandla, S. L.; Dale, B. E. Ammonia fiber explosion (AFEX) pretreatment of lignocellulosic wastes. American Institute of Chemical Engineers National Meeting, Chicago, IL, 1990.

Holtzapple, M. T.; Lundeen, J. E.; Sturgis, R.; Lewis, J. E.; Dale, B. E. *Appl. Biochem. Biotechnol.* 1992a, 34, 5-21.

Holtzapple, M. T.; Davison, R. R.; Stuart, E. D. 1992b. *US patent* 5, 171, 592.

Holtzapple, M. T. In *Encyclopedia of Food Science Food Technology and Nutrition;* Macrae, R.; Robinson, R. K.; Sadler, M. J.; Eds.; vol. 2. Academic Press: London, 1993, pp 2324-2334.

Horton, G. L.; Rivers, D. B.; Emert, G. H. *Ind. Eng. Chern. Prod. Res. Dev.* 1980, 79, 422-429.

Howell, J. A.; Stuck, J. D. *Biotechnol. Bioeng.* 1975, 17, 873-893.

Hsu, T. A.; Ladisch, M. R.; Tsao, G. T. *Chem. Technol.* 1980, 10, 315-319.

Hsu, T. -A. In *Handbook on Bioethanol, Production and Utilization*; Wyman, C. E.; Ed.; Taylor and Francis: Washington, DC, 1996, pp 179-212.

Hwang, S. S.; Lee, S. J.; Kim, H. K.; Ka, J. O.; Kim, K. J.; Song, H. G. *J. Microbiol. Biotechnol.* 2008, 18, 1819-1825.

Imai, M.; Ikari, K.; Suzuki, I. *Biochem. Eng. J.* 2004, 17, 79-83.

Ingram, L. O.; Doran, J. *FEMS Microbiol. Rev.* 1995, 16, 235-241.

Ingram, L. O.; Gomez, P. F.; Lai, X.; Moniruzzaman, M.; Wood, B. E.; Yomano, L. P.; York, S. W. Metabolic engineering of bacteria for ethanol production. *Biotechnol. Bioeng.* 1998, 58, 204-214.

Ingram, L. O.; Aldrich, H. C.; Borges, A. C. C.; Causey, T. B.; Martinez, A.; Morales, F.; Saleh, A.; Underwood, S.A.; Yomano, L. P.; York, S. W.; Zaldivar, J.; Zhou, S. D. *Biotechnol. Progr.* 1999, 15, 855-866.

Irwin, D.; Spezio, M.; Walker, L. P.; Wilson, D. B. *Biotechnol. Bioeng.* 1993, 42, 1002-1013.

Ishizawa, C. I.; Davis, M. F.; Schell, D. F.; Johnson, D. K. *J. Agric. Food. Chem.* 2007, 55, 2575-2581.

Israilides, C. J.; Grant, G. A.; Han, Y. W. *Appl. Environ. Microbiol.* 1978, 36, 43-46.

Iyer, P. V.; Wu, Z. -W.; Kim, S. B.; Lee, Y. Y. *Appl. Biochem. Biotechnol.* 1996, 57/58, 121-132.

Jeffries, T. W.; Shi, N. *Adv. Biochem. Eng./Biotechnol.* 1999, 65, 118-161.

Jeoh, T.; Ishizawa, C. I.; Davis, M. F.; Himmel, M. E.; Adney, W. S.; Johnson, D. K. *Biotechnol. Bioeng.* 2007, 98, 112-122.

Jimenez, L.; Bonilla, J. L. *Process Biochem.* 1993, 28, 243-247.

Kaar, W. E.; Holtzapple, M. T. *Biomass Bioenergy* 2000, 18, 189-199.

Kadam, K. L.; Camobreco, V. J.; Glazebrook, B. E.; Forrest. L. H.; Jacobson, W. A.; Simeroth, D. C.; Blackburn, W. J.; Nehoda, K. C. *Environmental Life Cycle Implications of Fuel Oxygenates Production from California Biomass*, NREL/TP-580-25688; National Renewable Energy Laboratory: 1999, Golden, CO.

Karimi, K.; Kheradmandinia, S.; Taherzadeh, M. J. *Biomass Bioenergy*, 2006, 30, 247-253.

Katzen, R.; Madson, P.W.; Monceaux, D.A. In The *Alcohols Textbook*; Lyons, T. P.; Murtagh, J. E.; Kelsall, D. R.; Eds.; Notingham University Press, 1995, pp 37-46.

Keller, F. A.; Hamilton, J. E.; Nguyen, Q. A. *Appl. Biochem. Biotechnol.* 2003, 105, 27-41.

Kelsey, R. G.; Shafizadeh, F. *Biotech. Bioeng.* 1980, 22, 1025-1236.

Kerem, Z.; Friesem, D.; Hadar, Y. *Appl. Environ. Microbiol.* 1992, 58, 1121-1127.

Khan, A. W.; Labrie, J. P., Mckeown, *J. Biotechnol. Bioeng.* 1986, 28, 1449-1453.

Khan, A.W.; Labrie, J.; McKeown, *J. Rad. Phys. Chem.* 1987, 29, 117-120.

Kilzer, F. J.; Broido, A. *Pyronamics.* 1965, 2, 151-163.

Kim, D. W.; Kim, T. S.; Jeong, Y. K.; Lee, J. K. J. *Ferment. Bioeng.* 1992, 73, 461-466.

Kim, J. S.; Lee, Y. Y.; Park, S. C. *Appl. Biochem. Biotechnol.* 2000, 84/86, 129-139.

Kim, S. B.; Um, B. H.; Park, S. C. Effect of pretreatment reagent and hydrogen peroxide on enzymatic hydrolysis of oak in percolation process. *Appl. Biochem. Biotechnol.* 2001, 91, 81-94.

Kim, T. H.; Kim, J. S.; Sunwoo, C.; Lee, Y. Y. In *24th Symposium on Biotechnology for Fuels and Chemicals*, 2002.

Kim, T. H.; Kim, J. S.; Sunwoo, C.; Lee, Y. Y. *Bioresour. Technol.* 2003, 90, 39-47.

Kim, K. H. *J. Sci. Food Agric.* 2005, 85, 2461-2467.

Kim, S.; Holtzapple, M. T. *Bioresour. Technol.* 2005, 96, 1994-2006.

Kim, S.; Holtzapple, M. T. *Bioresour. Technol.* 2006, 97, 583-591.

Kirk, T. K.; Tien, M.; Johnsrud, S. C.; Eriksson, K. E. *Enzyme Microb. Technol.* 1986, 8, 75-80.

Kirk, T. K.; Farrell, R. L. *Annu. Rev. Microbiol.* 198, 41, 465-505.

Kitchaiya, P.; Intanakul, P.; Krairiksh, M. *J. Wood Chem. Technol.* 2003, 23, 217-225.

Kleman-Leyer, K.; Agosin, E.; Conner, A. H.; Kirk, T. K. *Appl. Environ. Microbiol.* 1992, 58, 1266-1270.

Kleman-Leyer, K. M.; Gilkes, N. R.; Miller, R. C. Jr.; Kirk, T. K. *Biochem. J.* 1994, 302, 463 -469.

Kleman-Leyer, K. M.; Siika-Aho, M.; Teeri, T. T.; Kirk, T. K. *Appl. Environ. Microbiol.* 1996, 62, 2883-2887.

Klemm, D.; Philipp, B.; Heinze, T.; Heinze, U.; Wagenknecht, W. Comprehensive cellulose chemistry. I. *Fundamentals and analytical methods*. Wiley-VCH: Weinheim, 1998.

Klinke, H. B.; Ahring, B. K.; Schmidt, A. S.; Thomsen, A. B. *Bioresour. Technol.* 2002, 82, 15-26.

Knappert, D.; Grethlein, H.; Converse, A. *Biotechnol. Bioeng.* 1980, XXII, 1449-1463.

Knappert, D.; Grethlein, H.; Converse, A. *Biotechnol. Bioeng.* 1981, 11, 67-77.

Kobayashi, T.; Sakai, Y.; Iizuka, K. *Bull. Agric. Chem. Soc. Jpn.* 1960, 24, 443-449.

Kohlmann, K.; Westgate, P. J.; Weil, J.; Ladisch, M. R. *Biological based systems for waste processing.* SAE paper # 932251, Society of Automotive Engineerings, Inc.: Warrendale, PA, 1993.

Kohlmann, K. L.; Sarikaya, A.; Westgate, P. J.; Weil, J.; Velayudhan, A.; Hendrickson, R.; Ladisch, M. R. In *Enzymatic Degradation of Insoluble Carbohydrates*; Saddler, J. N.; Penner, M. H.; Eds.; ACS Publishing, 1995, pp 237-255.

Kong, F.; Engler, C. R.; Soltes, E. *J. Appl. Biochem. Biotechnol.* 1992, 34/35, 23-35.

Kosan, B., Michels, C., Meister, F., 2008. Dissolution and forming of cellulose with ionic liquids. *Cellulose*, 15, 59-66.

Koullas, D. P.; Christakopoulos, P. F.; Kekos, D.; Macris, B. J.; Koukios, E. G. *Cellulose Chem. Technol.* 1990, 24, 469-474.

Koullas, D. P.; Christakopoulos, P. F.; Kekos, D.; Koukios, E. G.; Macris, B. *J. Biomass Bioenergy,* 1993, 4, 9-13.

Kruus, K.; Wang W. K.; Ching, J.; Wu J. H. D. *J. Bacteriol.* 1995, 177, 1641-1644.

Kubikova, J.; Lu, P.; Zemann, A.; Krkoska, P.; Bobleter, O. *Cellulose Chem. Technol.* 2000, 34, 151-162.

Kuhar, S.; Nair, L. M.; Kuhad, R. C. *Can. J. Microbiol.* 2008, 54, 305-313.

Kumakura, M.; Kaetsu, I. *Biotech. Bioeng.* 1978, 20, 1309-1315.

Kumakura, M.; Kaetsu, I. *Int. J. Appl. Radiat. Isot.* 1979, 30, 139-141.

Kumakura, M.; Kaetsu, I. *Radiat. Phys. Chem.* 1984, 23, 523-527.

Kuo, C. H.; Lee, C. K. *Bioresour. Technol.* 2009, 100, 866-871.

Kurakake, M.; Ide, N.; Komaki, T.*Curr. Microbiol.* 2007, 54, 424-428.

Ladisch, M. R.; Dyck, K. *Science* 1979, 205, 898-900.

Ladisch, M. R.; Hong, J.; Voloch, M.; Tsao, G. T. In *Trends in the biology of fermentations for fuels and chemicals;* Hollaender, A.; Rabson, R.; Rodgers, P.; San Pietro, A.; Valentine, R.; Wolfe, R.; Eds.; Plenum Publishing: New York, 1981, pp 55-83.

Ladisch, M. R.; Lin, K. W.; Voloch, M.; Tsao, G. T. *Enzyme Microb.Technol.* 1983, 5, 82-102

Ladisch, M. R.; Voloch, M.; Hong, J.; Bienkowski, P.; Tsao, G. T. I&EC Process Des. Dev. 1984, 23, 437-443.

Lai, Y. -Z. In *Wood and cellulose chemistry*; Hon, D. N. -S.; Shiraishi, N.; Eds.; Marcel Dekker, New York, USA, 1991.

Laser, M.; Schulman, D.; Allen, S. G.; Lichwa, J.; Antal, M. J.; Lynd, L. R. *Bioresour. Technol.* 2002, 81, 33-44.

Laureano, -P. L.; Teymouri, F.; Alizadeh, H.; Dale, B. E. *Appl. Biochem. Biotechnol.* 2005, 121, 1081-1099.

Lawton, E. J.; Bellamy, W. D.; Hungate, R. E.; Bryant, M. P.; Hall, E. *Science*, 1951, 113, 380-382.

Lee, S. B.; Shin, H. S.; Ryu, D. D. Y. *Biolectmol. Bioeng*. 1982, 24, 2137-53.

Lee, Y. -H.; Fan, L. T. *Biotechnol. Bioeng*. 1982, 24: 2383-2406.

Lee, S. B., Kim, I. H.; Ryu, S. S. Y.; Taguchi, H. *Biotechnol. Bioeng*. 1983, 25, 33-51.

Lee, D.; Yu, A. H. C.; Wong, K. K. Y.; Saddler, J. R. *Appl. Biochem. Biotechnol*. 1994, 45/45, 407-415.

Lee, J. W.; Gwak, K. S.; Park, J. Y.; Park, M. J.; Choi, D. H.; Kwon, M.; Choi, I. G. *J. Microbiol*. 2007, 45, 485-491.

Lenze, J.; Esterbauer, H.; Sattler, W.; Schurz, J.; Wrentschur, E. *J. Appl. Polym. Sci*. 1990, 41, 1315-1326.

Li, Y.; Ruan, R.; Chen, P. L.; Liu, Z.; Pan, X.; Lin, X.; Liu, Y.; Mok, C. K.; Yang, T. *Trans. ASABE,* 2004, 47, 821-825.

Li, A.; Antizar-Ladislao, B.; Khraisheh, M. *Bioprocess. Biosyst. Eng*. 2007, 30, 189-196.

Lin, K. W.; Ladisch, M. R.; Schaefer, D. M.; Noller, C. H.; Lechtenberg, V.; Tsao, G. T. *AIChE Symp*. 1981, 207, 102-106.

Linden, J. C.; Decker, S. R.; Samara, M. *ACS Symp. Ser*. 1994; 566, 452--467.

Liu, C. G.; Wyman, C. E. *Appl. Biochem. Biotechnol*. 2004, 113/116, 977-987.

Liu, L. Y.; Chen, H. Z. *Chin. Sci. Bull*. 2006, 51, 2432-2436.

Lora, J. H.; Pye, E. K. In *Tappi Notes-1992 Solvent Pulping Symp*. 1992, pp 27-34.

Lu, Z. X.; Kumakura, M. *Isotopes* Environ *Health* Stud. 1995, 31, 151-160.

Lynd, L. R.; Cushman, J. H.; Nichols, R. J.; Wyman, C. E. *Science* 1991, 251, 1318-1323.

Lynd, L. R. *Annu. Rev. Energy Environ*. 1996, 21, 403-65.

Lynd, L. R.; Elander, R.T.; Wyman, C. E. *Appl. Biochem. Biotechnol*. 1996, 57/58, 741-761.

Lynd, L. R., Weimer, P. J.; van Zyl, W. H.; Pretorius, I. S. *Microbiol. Mol. Biol. Rev*. 2002, 66, 506-577.

Lynd, L. R.; Van Zyl, W. H.; McBride, J. E.; Laser, M. *Curr. Opin. Biotechnol*. 2005, 16, 577-583.

Ma, H.; Liu, W. W.; Chen, X., Wu, Y. J., Yu, Z. L. *Bioresour. Technol*. 2009, 100, 1279-1284.

Mackie, K. L.; Brownell, H. H., West, K. L.; Saddler, J. N. *J. Wood Chem. Technol*. 1985, 5, 405-425.

Mahinpour, R.; Sarbolouki, M. N. Iran. *J. Chem. Chem. Eng. -Int. English Ed*. 1998, 17, 8-13.

Mais, U.; Esteghlalian, A. R.; Saddler, J. N.; Mansfield S. D. *Appl. Biochem. Biotechnol*. 2002, 98, 815-832.

Maloney, T.; Rantanen, J.; Paulapuro, H.; Yliruusi, *J. Pharm. Res*. 2003, 18, 1562.

Mamers, H. D.; Menz, D. N. *J. Appita*. 1984, 37, 644-649.

Mandels, M.; Kostick, J.; Parizek, R. *J. Polymer. Sci*. 1971, 36, 445-459.

Mandels, M.; Hontz, L.; Nystrom, *J. Biotechnol. Bioeng*. 1974, 16, 1471-1493.

Mansfield, S. D.; de Jong, E.; Stephens, R. S.; Saddler, J. N. *J. Biotechnol*. 1997, 57, 205.

Mansfield, S. D.; Mooney, C.; Saddler, J. N. *Biotechnol. Prog*. 1999, 15, 804-816.

Marshall, K.; Sixsmith, D. *Drug Dev. Comm*. 1974, 1, 51-71.

Martin, C.; Thomsen, A. B. J. *Chem. Technol. Biotechnol*. 2007, 82, 174-181.

Martin, C.; Alriksson, B.; Sjode, A.; Nilvebrant, N. O.; Jonsson, L. *J. Appl. Biochem. Biotechnol*. 2007a, 137, 339-352.

Martin, C.; Klinke, H. B.; Thomsen, A. B. *Enzyme Microb. Technol*. 2007b, 40, 426-432.

Martin, C.; Thomsen, M. H.; Hauggaard-Nielsen, H.; BelindaThomsen, A. *Bioresour Technol*. 2008, 99, 8777-8782.

Martinez, J. M.; Reguant, J.; Montero, M. A.; Montane, D.; Salvado, J.; Farriol, X. *Ind. Eng. Chem. Res*. 1997, 36, 688-696.

Marzialetti, T.; Olarte, M. B. V.; Sievers, C.; Hoskins, T. J. C.; Agrawal, P. K.; Jones, C. W. *Ind. Eng. Chem. Res*. 2008, 47, 7131-7140.

Mason, W. H. 1926. *US Patent* 1,578,609.

Mcdonald, A. G.; Clark, T. A. *J. Wood Chem. Technol*. 1992, 12, 53-78.

McGinnis, G. D.; Wilson, W. W.; Muller, C. E. *Ind. Chem. Prod. Res. Dev*. 1983, 22, 352–357.

McKendry, P. *Bioresour. Technol*. 2002, 83, 37-46.

McMillan, J. D. In *Enzymatic Conversion of Biomass for Fuels Production*; Himmel, M. E.; Baker, J. O.; Overend, R. P.; Eds.; ACS Symp. Ser.; ACS, Washington, DC, 1994; vol. 566, pp 292-324.

Mehlberg, R. L. 1977. M.S. Thesis, Purdue University, West Lafayette, IN. Mehlberg, R. L.; Tsao, G. T. Presented at the 178th ACS National Meeting, Washington, DC, 1979.

Mes-Hartree, M.; Dale, B. E.; Craig, W. K. *Appl. Microbiol. Biotechnol*. 1988, 29, 462-468.

Michalowicz, G.; Toussaint, B.; Vignon, M. R. *Holzforschung*, 1991, 45, 175-179.

Mikkola, J. P.; Kirilin, A.; Tuuf, J. C.; Pranovich, A.; Holmbom, B.; Kustov, L. M.; Murzin, D. Y.; Salmi, T. Green Chem. 2007, 9, 1229-1237.

Millet, M. A.; Baker, A. J.; Satter, L. D. *Biotech. Bioeng. Symp*. 1976, 6, 125-153.

Mishima, D.; Tateda, M.; Ike, M.; Fujita, M. *Bioresour. Technol*. 2006, 97, 2166-2172.

Mohagheghi, A.; Evans, K.; Finkelstein, M.; Zhang, M. *Appl. Biochem. Biotechnol*. 1998, 70/72, 285-299.

Mok, W. S. -L.; Antal, Jr. M. J. *Ind. Eng. Chem. Res*. 1992, 31, 1157-1161.

Mok, W. S. -L.; Antal, Jr. M. J. In *Advances in Thermochemical Biomass Conversion*; Bridgewater, A. V.; Ed.; Blackie Academic and Professional Publishers, New York, 1994; vol. 2, pp 1572-1582.

Mooney, C. A.; Mansfield, S. D.; Touhy, M. G.; Saddler, J. N. *Bioresour. Technol*. 1998, 64, 113-119.

Mooney, C. M.; Mansfield, S. D.; Beatson, R. P.; Saddler, J. N. *Enzyme Microb. Technol*. 1999, 25, 644-650.

Morjanoff, P. J.; Gray, P. P. *Biotechnol. Bioeng*. 1987, 29, 733-741.

Mosier, N. S.; Hendrickson, R.; Dreschel R.; Welch, G.; Dien, B. S.; Bothast, R.; Ladisch, M. R. *Principles and economics of pretreating cellulose in water for ethanol production*, Paper No. 103, BIOT Division, 225th American Chemical Society Meeting, New Orleans, 2003a, March.

Mosier, N. S.; Hendrickson, R.; Welch, G.; Dreschel, R.; Dien, B.; Ladisch, M. R. Corn fiber pretreatment scale-up and evaluation in an industrial corn to ethanol facility. Paper No. 6A-04, 25th Symposium on Biotechnology for Fuels and Chemicals, Breckenridge, CO, 2003b.

Mosier, N.; Wyman, C.; Dale, B.; Elander, R.; Lee, Y. Y.; Holtzapple, M.; Ladisch, M. *Bioresour. Technol*. 2005a, 96, 673-686.

Mosier, N. S.; Hendrickson, R.; Brewer, M.; Ho, N.; Sedlak, M.; Dreshel, R.; Welch, G.; Dien, B. S.; Aden, A.; Ladisch, M. R. *Appl. Biochem. Biotechnol*. 2005b, 125, 77-97.

Mosier, N.; Hendrickson, R.; Ho, N.; Sedlak, M.; Ladisch, M. R. *Bioresour. Technol.* 2005c, 96, 1986-1993.

Myerly, R. C.; Nicholson, M. D.; Katzen, R.; Taylor, J. *M. Chemtech.* 1981, 11, 186-192.

Nahzad, M. M.; Ramos, L. P.; Paszner, L.; Saddler, J. N. *Enzyme Microb. Technol.* 1995, 17, 68.

National Research Council. (1999). Committee on Biobased Industrial Products, Biobased Industrial Products-Priorities for Research and Commercialization. National Academy Press.

Negro, M. J.; Manzanares, P.; Oliva, J. M.; Ballesteros, I.; Ballesteros, M. *Biomass Bioenergy* 2003a, 25, 301-308.

Negro, M. J.; Manzanares, P.; Ballesteros, I.; Oliva, J. M.; Cabanas, A.; Ballesteros, M. Appl. *Biochem. Biotechnol.* 2003b, 105, 87-100.

Neilson, M. J.; Kelsey, R. G.; Shafizadeh, F. *Biotech. Bioeng.* 1982, 24, 293-304.

Neuman, R. P.; Walker, L. P. *Biotechnol. Bioeng.* 1992, 40, 226-234.

Neureiter, M.; Danner, H.; Fruhauf, S.; Kromus, S.; Thomasser, C.; Braun, R.; Narodoslawsky, M. *Bioresour. Technol.* 2004, 92, 21-29.

Nguyen, Q. A.; Tucker, M. P.; Keller, F. A.; Beaty, D. A.; Connors, K. M.; Eddy, F. P. *Appl. Biochem. Biotechnol.* 1999, 77-79, 133-142.

Nguyen, Q. A.; Tucker, M. P.; Keller, F. A.; Eddy, F. P. *Appl. Biochem. Biotechnol.* 2000, 84/86, 561-576.

Nitayavardhana, S.; Rakshit, S. K.; Grewell, D.; van Leeuwen, J. H.; Khanal, S. H. *Biotechnol. Bioeng.* 2008, 101, 487-496.

Ohgren, K.; Bura, R.; Saddler, J.; Zacchi, G. *Bioresour. Technol.* 2007, 98, 2503-2510.

Ohmine, K.; Ooshima, H.; Harano, Y. *Biotechnol. Bioeng.* 1983, 25, 2041-2053.

Oksanen T, Pere J, Paavlainen L, Buchert J, Viikari L (2000) *J. Biotechnol* 78:39.

Onuki, S.; Koziel, K. A.; van Leeuwen, J.; Jenks, W.S.; Grewell, D.; Cai, L. S. *ASABE,* paper No. 085136, Providence, RI, 2008.

Ooshima, H.; Sakata, M.; Harano, Y. *Biotechnol. Bioeng.* 1983, 25, 3103-3114.

Ooshima, H.; Aso, K.; Harano, Y.; Yamamoto, T. *Biotechnol. Lett.* 1984, 6, 289-294.

OpTest Equipment. (1999). *HiRes fiber quality analyzer-user manual.* OpTest, Washington DC, USA.

Palmqvist, E.; Hahn-Hagerdal, B. *Bioresour. Technol.* 2000, 74, 25-33.

Pan, X.; Zhang, X.; Gregg, D. J.; Saddler, J. N. *Appl. Biochem. Biotechnol.* 2004, 113-116, 1103-1114.

Pan, X.; Zhang, X.; Xie, D.; Gregg, D. J.; Saddler, J. N. *Appl. Biochem. Biotechnol.* 2005, 121/124, 1069-1080.

Pan, X. J.; Gilkes, N.; Kadla, J.; Pye, K.; Saka, S.; Gregg, D.; Ehara, K.; Xie, D.; Lam, D.; Saddler, *J. Biotechnol. Bioeng.* 2006, 94, 851-861.

Pan, X.; Xie, D.; Kang, K-. Y.; Yoon, S-. L.; Saddler, J. N. *Appl. Biochem. Biotechnol.*, 2007, 136/140, 367-377.

Pan, X.; Xie, D.; Yu, Y. W.; Saddler, J. N. *Biotechnol. Bioeng.* 2008, 101, 39-48.

Palonen, H.; Thomsen, A. B.; Tenkanen, M.; Schmidt, A. S.; Viikari, U. *Appl. Biochem. Biotechnol.* 2004, 117, 1-17.

Patel, M. M.; Bhatt, R. M. *J. Chem. Technol. Biotechnol.* 1992, 53, 253-263.

Pereira, A. N.; Mobedshahi, M.; Ladisch, M. R. *Methods Enzymol.* 1988, 160, 26-43.

Perez, J. A.; Gonzalez, A.; Oliva, J. M.; Ballesteros, I.; Manzanares, P. *J. Chem. Technol. Biotechnol.* 2007, 82, 929-938.

Perez, J.A.; Ballesteros, I.; Ballesteros, M.; Saez, F.; Negro, M. J.; Manzanares, P. *Fuel*, 2008, 87, 3640-3647.

Pinto, J. H.; Kamden, D. P. *Appl. Biochem. Biotechnol.* 1996, 61, 289-297.

Playne, M. J. *Biotechnol. Bioeng.* 1984, 26, 426-433.

Prior, B. A.; Day, D. F. *Appl. Biochem. Biotechnol.* 2008, 146, 151-164.

Pritchard, G. I.; Pigden, W. J.; Minson, D. *J. Can. J. Anita. Sci.* 1962, 42, 215.

Puls, J.; Wood, T. M. *Bioresour. Technol.* 1991, 36, 15-19.

Puri, V. P.; Mamers, H. *Biotech. Bioeng.* 1983, 25, 3149-3161.

Puri, V. P. *Biotechnol. Bioeng.* 1984, 26, 1219-1222.

Purwadi, R. 2006. Continuous Ethanol Production from Dilute-acid Hydrolyzates: Detoxification and Fermentation Strategy. PhD Thesis, Chalmers University Of Technology, Göteborg, Sweden.

Ramos, L. P.; Breuil, C.; Saddler, J. N. *Appl. Biochem. Biotechnol.* 1992, 34/35, 37-48.

Ramos, L. P.; Nazhad, M. M.; Saddler, J. N. *Enzyme Microb. Technol.* 1993, 15, 821-831.

Reinikainen T.; Teleman, O.; Teeri, T. T. *Proteins* 1995, 22, 392-403.

Reverbel-Leroy, C.; Pages, S.; Belaich, A.; Belaich. J. P.; Tardif, C. *J. Bacteriol.* 1997, 179, 46-52.

Rivers, D. B.; Emert, G. H. *Biotech. Lett.* 1987, 9, 365-368.

Rivers, D. B.; Emert, G. H. *Biotech. Bioeng.* 1988, 31, 278-281.

Rogalinski, T.; Ingram, T.; Brunner, G. J. *Supercrit. Fluids* 2008, 47, 54-63.

Rogers, R. D. *Chem. Eng. News* 2002, 80, 4-5.

Rogers, R. D.; Holbrey, J. D.; Spear, S. K.; Turner, M. B. *Abstracts of Papers of the ACS*, 2004, 227, U310-U310.

Ropare, M.; Marchal, R.; Pourquie, J.; Vandecasteele, J. P. *Bioresour. Technol.* 1992, 42, 197-204.

Rotter, B. E.; Barry, D. A.; Gerhard, J. I.; Small, J. S. *Bioresour. Technol.* 2008, 99, 5738-5748.

Ruiz, E.; Cara, C.; Manzanares, P.; Ballesteros, M.; Castro, E. *Enzyme Microb. Technol.* 2008, 42, 160-166.

Ruohonen, L.; Aristidou, A.; Frey, A. D.; Penttila, M.; Kallio, P. T. *Enzyme Microb. Technol.* 2006, 39, 6-14.

Rydholm, S. A. *Pulping Processes. Interscience Publishers*, John Wiley and Sons; New York, 1965; pp 3-90.

Ryu, D. D.; Lee, S. B.; Tassinari, T.; Macy, C. *Biotechnol. Bioeng.* 1982, 24, 1047-1067.

Ryu, S. K.; Lee, J. M. *Biotechnol. Bioeng.* 1983, 25, 53-65.

Saddler, J. N.; Brownell, H. H.; Clermont, L. P.; Levitin, N. *Biotechnol Bioeng.* 1982. 24, 1389-1402.

Saddler, J. N.; Mes-Hartree, M.; Yu, E. K. C.; Brownell, H. H. *Biotech. Bioeng. Symp.* 1983, 13, 225-238.

Saddler, J. N.; Ramos, L. P.; Breuil, C. In *Bioconversion of Forest and Agricultural Plant Wastes*; Saddler, J. N.; Ed.; C.A.B. International, Wallingford, UK, 1993; pp 73-92.

Saha, B. C. *Biotechnol. Adv.* 2000, 18, 403-423.

Saha, B. C.; Iten, L. B.; Cotta, M. A.; Wu, Y. V. *Process Biochem.* 2005, 40, 3693-3700.

Saha, B. C.; Cotta, M. A. *Biotechnol. Progr.* 2006, 22, 449-453.

Saha, B. C.; Cotta, M. A. *Enzyme Microb. Technol.* 2007, 41, 528-532.

Saha, B. C.; Biswas, A.; Cotta, M. A. *J. Biobased Mater. Bioenergy* 2008, 2, 210-217.

Salvesen, J. R.; Brink, D. L.; Diddams, D. G.; Owzarskl, P.; Vanlllln. 1952. *US Patent* 2,434,626,

Sarkanen, K. V. In *The chemistry of wood; Browning*, B. L.; Ed.; Robert E., Krieger Publishing Company; New York, 1975; Chapter 6.

Sarkanen, K.V. *Prog. Biomass Convers.* 1980, 2, 127-144.

Sakai, Y. *Bull. Chem. Soc. Jpn.* 1965, 38, 863-868.

Sasaki, T.; Tanaka, T.; Nanbu, N.; Sato, Y.; Kainuma, K. *Biotechnol. Bioeng.* 1979, 21, 1031-1042.

Schell, D. J.; Farmer, J.; Newman, M.; McMillan, J. D. *Appl. Biochem. Biotechnol.* 2003. 105, 69-85.

Schmid, G.; Bisell, M.; Wandrey, C. *Anal. Biochem.* 1988, 175, 573-583.

Schmidt, A. S.; Thomsen, A. B. *Bioresour. Technol.* 1998, 64, 139-151.

Schurz, J. In *Bioconversion of Cellulosic Substances into Energy Chemicals and Microbial Protein Symp.* Proc.; Ghose, T. K.; Ed.; New Delhi, IIT: Delhi, 1978; pp 37.

Schurz, J.; Billiani, J.; Honel, A.; Eigner, W. D.; Ja´nosi, A.; Hayn, M.; Esterbauer, H. *Acta Polym.* 1985, 36, 76-80.

Schurz, J. Holzforschung 1986, 40, 225-232.

Schwald, W.; Breuil, C.; Brownell, H. H.; Chan, M.; Saddler, J. N. *Appl. Biochem. Biotechnol.* 1989a, 20/21, 29-44.

Schwald, W.; Smaridge, T.; Chan, M.; Breuil, C.; Saddler, J. N. In *enzyme Systems for Lignocellulose Degradation*; Coughlan, M. P.; Ed.; Elsevier: New York, 1989b; pp 231-242.

Selby, K. *Biochem. J.* 1961, 79, 562-566.

Shafizadeh, F.; Lai, Y. -Z. *Carbohyd. Res.* 1975, 42, 39-53.

Shafizadeh, F. (1977). *TAPPI Forest Biology Wood Chemistry Conf.*, Madison, June 20-22.

Shafizadeh, F.; Bradbury, A. G. W.; Sakai, Y. *Abstracts of Papers of the ACS*, 1978, 176, 79-79.

Shafizadeh, F.; Bradbury, A. G. W. *J. Appl. Polym.* Sci. 1979, 23, 1431-1442.

Sheehan, J., and Himmel, M. E. (1999). *Biotechnology Progress*, 15, 817–827.

Shevchenko, S. M.; Chang, K.; Robinson, J.; Saddler, J. N. *Bioresour. Technol.* 2000, 72, 207.

Shi, J.; Chinn, M. S.; Sharma-Shivappa, R. R. *Bioresour. Technol.* 2008, 99, 6556-6564.

Shimizu, K. *Mokuzai Bakkaishi* 1980, 26, 488-493.

Shimizu, K. *Jpn. Agr. Res.* Q. 1981, 14, 424-4248.

Shin, S. J.; Sung, Y. J. *Radiat. Phys. Chem.* 2008, 77, 1034-1038.

Sidiras, D.; Koukios, E. *Bioresource Technol.* 2004, 94, 91-98.

Silverstein, R. A.; Chen, Y.; Sharma-Shivappa, R. R.; Boyette, M. D.; Osborne, *J. Bioresour. Technol.* 2007, 98, 3000-3011.

Singh, P.; Suman, A.; Tiwari, P.; Arya, N.; Gaur, A.; Shrivastava, A. K. *World J. Microbiol. Biotechnol.* 2008, 24, 667-673.

Sinitsyn, A. P.; Mitkevich, O. V.; Gusakov, A, V. *Carbohydr. Polym.* 1989, 10, 1-14.

Sinitsyn, A. P.; Gusakov, A. V.; Vlasenko, E. Y. *Appl. Biochem. Biotechnol.* 1991, 30, 43-59.

Sivers, M.V.; Zacchi, G. *Bioresour. Technol.* 1995, 51, 43-52.

Sjostrom, E. *Wood chemistry-fundamentals and applications 2nd edn*; Academic Press; Washington DC, USA, 1981.

Sjostrom, E., *Wood chemistry-Fundamentals and Applications*; Academic Press; New York, 1985.

Soderstrom, J.; Pilcher, L.; Galbe, M.; Zacchi, G. *Appl. Biochem. Biotechnol.* 2002, 98, 5-21.

Soderstrom, J.; Pilcher, L.; Galbe, M.; Zacchi, G. *Appl. Biochem. Biotechnol.* 2003a, 105, 127-140.

Soderstrom, J.; Pilcher, L.; Galbe, M.; Zacchi, G. *Biomass Bioenergy* 2003b, 24, 475-486.

Sorensen, A.; Teller, P. J.; Hilstrom, T.; Ahring, B. K. *Bioresour. Technol.* 2008, 99, 6602-6607.

Srisodsuk, M.; Kleman-Leyer, K.; Keranen, S.; Kirk, T. K.; Teeri, T. T. *Eur. J. Chem.* 1998, 251, 885-892.

Stenberg, K.; Tengborg, C.; Galbe, M.; Zacchi, G. *J. Chem. Technol. Biotechnol.* 1999, 71, 299-308.

Stone, J. E.; Scallan, A. M.; Donefer, E.; Ahlgren, E. In *Cellulases and their Application*; Gould, R. E.; Ed.; ACS: Washington DC, 1969; pp 219-241.

Sues, A.; Millati, R.; Edebo, L.; Taherzadeh, M. J. *FEMS Yeast Res.* 2005, 5, 669-676.

Sun, Y.; Cheng, J. *J. Bioresour. Technol.* 2002, 83, 1-11.

Sun, Y.; Cheng, J. *J. Bioresour. Technol.* 2005, 96, 1599-1606.

Sun, F.; Chen, H. *Bioresour. Technol.* 2008, 99, 6156-6161.

Swatloski, R. P.; Spear, S. K.; Holbrey, J. D.; Rogers, R. D. *J. Am. Chem. Soc.* 2002, 124, 4974-4975.

Swatloski, R. P.; Spear, S. K.; Holbrey, J. D.; Rogers, R. D. *Abstracts of Papers of the ACS*, 2003, 225, U288-U288.

Taherzadeh, M. J.; Karimi, K. *Int. J. Mol. Sci.* 2008, 9, 1621-1651.

Tanahashi, M.; Goto, T.; Horii, F.; Hirai, A.; Higuchi, T. Mokuzai Gakkaishi 1989, 35, 654-662.

Tanaka, M.; Ikesaka, M.; Matsuno, R. *Biotechnol. Bioeng.* 1988, 32, 698-706.

Taniguchi, M.; Suzuki, H.; Watanabe, D.; Sakai, K.; Hoshino, K.; Tanaka, T. *J. Biosci. Bioeng.* 2005, 100, 637-643.

Tassinari, T.; Macy, C. *Biotech. Bioeng.* 1977, 19, 1321-1330.

Tassinari, T.; Macy, C.; Spano, L. *Biotech. Bioeng.* 1980, 22, 1689-1705.

Tassinari, T.; Macy, C.; Spano, L. *Biotech. Bioeng.* 1982, 24, 1495-1505.

Teeri, T. T. *Trends Biotechnol.* 1997, 15, 160-167.

Teixeira, L. C.; Linden, J. C.; Schroeder, H. A. *Appl. Biochem. Biotechnol.* 1999a, 77/79, 19-34.

Teixeira, L. C.; Linden, J. C.; Schroeder, H. A. *Renewable Energy* 1999b, 16, 1070-1073.

Teixeira, L.C.; Linden, J.C.; Schroeder, H.A. *Appl. Biochem. Biotechnol.* 2000, 84/86: 111-127.

Tengborg, C.; Stenberg, K.; Galbe, M.; Zacchi, G.; Larsson, S.; Palmqvist, E.; Hahn-Hagerdal, B. *Appl. Biochem. Biotechnol.* 1998, 70/72, 3-15.

Teramoto, Y.; Tanaka, N.; Lee, S-. H.; Endo, T. *Biotechnol. Bioeng.* 2008, 99, 75-85.

Thompson, D. N.; Chen, H.; Grethlein, H. E. In *Biotechnolgy in Pulp and Paper Manufacture*; Kirk, T. K.; Chang, H.; Eds.; Butterworth-Heinemann: Boston, 1990: pp 329-338.

Thompson, D. N.; Chen, H-. C.; Grethlein, H. E. *Bioresour. Technol.* 1992, 39, 155-163.

Thomsen, M. H.; Thygesen, A.; Thomsen, A. B. *Bioresour. Technol.* 2008, 99, 4221-4228.
Thring, R. W.; Chorent, E.; Overend, R. *Biomass* 1990, 23, 289-305.
Timell, T.E. *Wood Sci. Tech.* 1967, 1, 45-70.
Tokuyasu, K.; Tabuse, M.; Miyamoto, M.; Matsuki, J.; Yoza, K. *Carbohydr. Res.* 2008, 343, 1232-1236.
Torget, R.; Walter, P.; Himmel, M.; Grohmann, K. *Appl. Biochem. Biotechnol.* 1990, 24/25, 115-126.
Torget, R.; Walter, P.; Himmel, M.; Grohmann, K. *Appl. Biochem. Biotechnol.* 1991, 28/29, 75–86.
Torget, R.; Himmel, M.; Grohmann, K. *Appl. Biochem. Biotechnol.* 1992, 34/35, 115-123.
Torget, R.; Hsu, T. *Appl. Biochem. Biotechnol.* 1994, 45/46, 5-21.
Tsao, G. T.; Ladisch, M.; Ladisch, C.; Hsu, T. A.; Dale, B.; Chou, T. *Annu. Rep. Ferment. Processes* 1978, 2, 1-21.
Um, B. H.; Karim, M. N.; Henk, L. L. *Appl. Biochem. Biotechnol.* 2003, 105,115-125.
Valjamae, P.; Pettersson, G.; Johansson, G. *Eur. J. Biochem.* 2001, 268, 4520-4526.
Van Wyk, J. P. H. *Biotechnol. Tech.* 1997, 11, 443-445.
Van Wyk, J. P. H. *Trends Biotechnol.* 2001, 19, 172-177.
Van Walsum G. P.; Allen, S. G.; Spenser, M. J.; Laser, M. S.; Antal, M. J.; Lynd, L. R. *Appl. Biochem. Biotechnol.* 1996, 57/58, 157-170.
Varga, E.; Schmidt, A. S.; Reczey, K.; Thomsen, A. B. *Appl. Biochem. Biotechnol.* 2003, 104, 37-50.
Varga, E.; Reczey, K.; Zacchi, G. *Appl. Biochem. Biotechnol.* 2004, 120, 113.
Vazquez, M.; Oliva, M.; Tellez-Luis, S. J.; Ramirez, J. A. *Bioresour. Technol.* 2007, 98, 3053-3060.
Vedernikov, N.; Karlivans, V.; Roze, I.; Rolle, A. *Sibirskii Khimicheskii Zhurnal*, 1991, 5, 67-72.
Vlasenko, E.; Yu-Ding, H.; Labavitch, J. M.; Shoemaker, S. P. *Bioresour. Technol.* 1997, 57:109-119.
Walch, E.; Zemann, A.; Schinner, F.; Bonn, G.; Bobleter, O. *Bioresour. Technol.* 1992, 39, 173-177.
Waldner, R.; Leisola, M. S. A.; Fiechter, A. *Appl. Microbiol. Biotechnol.* 1988, 29, 400-407.
Walker, L. P.; Wilson, D. B. *Bioresour. Technol.* 1991, 36, 3-1 4.
Wayman, M.; Chen, S.; Doan, K. *Process Saf. Environ. Protect.* 1993, 71, 141-143.
Weil, J. R.; Sarikaya, A.; Rau, S. -L.; Goetz, J.; Ladisch, C. M.; Brewer, M.; Hendrickson, R.; Ladisch, M. R. *Appl. Biochem. Biotechnol.* 1997, 68, 21-40.
Weil, J. R.; Sarikaya, A.; Rau, S. -L.; Goetz, J.; Ladisch, C. M.; Brewer, M.; Hendrickson, R.; Ladisch, M. R. *Appl. Biochem. Biotechnol.* 1998a, 73, 1-17.
Weil, J. R.; Brewer, M.; Hendrickson, R.; Sarikaya, A.; Ladisch, M. R. *Appl. Biochem. Biotechno*l. 1998b, 70/72, 99-111.
Weimer, P. J.; Weston, W. J. *Biotechnol. Bioeng.* 1985, 27, 1540.
Weimer, P. J.; Lopez-Guisa, J. M.; French, A. D. *Appl. Environ. Microbiol.* 1990, 56, 2421-2419.
Weimer, P. J.; Hackney, J. M.; French, A. D. *Biotechnol. Bioeng.* 1995, 48, 169-178.
Wenzel, H. F. *The Chemical Technology of Wood.* Academic Press, 1970; pp 32-245.
Whitaker, D. R. *Can. J. Biochem. Physiol.* 1957, 35, 733-742.
White, A. R.; Brown, R. M. *Jr. Proc. Natl. Acad. Sci. USA* 1981, 78, 1047-1051.

Wilkie, K. C. B. *Adv. Carbohydr. Biochem.* 1979, 36, 215-264.

Wingren, A.; Galbe, M.; Roslander, C.; Rudolf, A.; Zacchi, G. *Appl. Biochem. Biotechnol.* 2005, 121, 485-499.

Wiselogel, A.; Tyson, S.; Johnson, D. In *Handbook on Bioethanol: Production and Utilization*; Wyman, C. E.; Ed.; Taylor and Francis: Washington, DC, 1996, pp 105-118.

Wojciak, A.; Pekarovicova, A. *Cellul. Chem. Technol.* 2001, 35, 361-369.

Wong, K. Y.; Deverell, K. F.; Mackie, K. L.; Clark, T. A. *Biotechnol. Bioeng.* 1988, 31, 447.

Wood, T. M. *Biotechnol. Bioeng. Symp.* 1975, 5, 111-137.

Wood, T. M.; McCrae, S. I. *Biochem. J.* 1978, 171, 61-72.

Wooley, R.; Ruth, M.; Glassner, D.; Sheehan, *J. Biotechnol. Progr.* 1999, 15, 794-803.

Wright, J. D.; Wyman, C. E.; Grohmann, K. *Appl. Biochem. Biotechnol.* 1987, 18, 75-90.

Wright, J. D. *Chem. Eng. Prog.* 1988, 84, 62-74.

Wu, J.; Zhang, J.; He, J.; Ren, Q.; Guo, M. *Biomacromolecules* 2004, 5, 266.

Wyman, M.; Chen, S.; Doan, K. *Tran. IChemE.* 1993, 71, 141-143.

Wyman, C. E. *Bioresour. Technol.* 1995a, 50, 3-15.

Wyman, C. E. In *Enzymatic Degradation of Insoluble Carbohydrates*; Saddler, J. N.; Penner, M. H.; Eds.; ACS Symp. Ser., ACS: Washington, DC, 1995b: vol. 618, pp 272-290.

Wyman, C. E. In Handbook on Bioethanol, Production and Utilization; Wyman, C. E.; Ed.; Taylor and Francis: Washington, DC, 1996, Chapter 1.

Wyman, C. E. *Annu. Rev. Energy Env.* 1999, 24, 189-226.

Wyman, C. E. *MRS Bull.* 2008, 33, 381-383.

Xin, L. Z.; Kumakura, M. *Bioresour. Technol.* 1993, 43, 13-17.

Yang, B.; Boussaid, A.; Mansfield, S. D.; Gregg, D. J.; Saddler, J. N. *Biotechnol. Bioeng.* 2002, 77, 678-685.

Yang, B.; Wyman, C. E. *Biotechnol. Bioeng.* 2004, 86, 88-98.

Yang, C. P.; Shen, Z. Q.; Yu, G. C.; Wang, *J. Bioresour. Technol.* 2008, 99, 6240-6245.

Yoon, H. H.; Wu, Z.; Lee, Y. Y. *Presented at the 16th Symposium on Biotechnology for Fuels and Chemicals*, Gatlinburg, TN, 1994, May 9-13.

Yoon, H. H.; Wu, Z. W.; Lee, Y. Y. *Appl. Biochem. Bio*technol. 1995, 51-52, 5-19.

Yoon, H. H. *Korean J. Chem. Eng.* 1998, 15, 631-636.

Yoshida, M.; Liu, Y.; Uchida, S.; Kawarada, K.; Ukagani, Y.; Ichinose, H.; Kaneko, S.; Fukuda, K. *Biosci. Biotechnol. Biochem.* 2008, 72, 805-810.

Youssef, B. M.; Aziz, N. H. *Cytobios* 1999, 97, 171-183.

Yu, J.; Zhang, J.B.; He, J.; Liu, Z. D.; Yu, Z. N. *Bioresour. Technol.* 2009, 100, 903-908.

Zeng, M.; Mosier, N. S.; Huang, C.-P.; Sherman, D. M.; Ladisch, M. R. *Biotech. Bioeng.* 2007, 97, 265-278.

Zhang, C.; Eddy, K.; Deanda, M. *Science,* 1995, 267, 240-243.

Zhang, Y. -H. P., Lynd, L. R. *Anal. Biochem.* 2003, 322: 225-232.

Zhang, Y. -H. P., Lynd, L. R. *Biotechnol. Bioeng.* 2004, 88, 797-824.

Zhang, H.; Wu, J.; Zhang, J.; He, *J. Macromolecules* 2005, 38, 8272-8277.

Zhang, W.; Liang, M.; Lu, C. *Cellulose* 2007a, 14, 447-456.

Zhang, X. Y.; Yu, H. B.; Huang, H. Y.; Liu, Y. X. *Int. Biodeterior. Biodegrad.* 2007b, 60, 159-164.

Zhang, X. Y.; Xu, C. Y.; Wang, H. X. *J. Biosci. Bioeng.* 2007c, 104,149-151.

Zhao, X. B.; Wang, L.; Liu, D. H. *J. Chem. Technol. Biotechnol.* 2007, 82, 1115-1121.

Zhao, X. B.; Wang, L.; Liu, D. H. *J. Chem. Technol. Biotechnol.* 2008a, 83, 950-956.

Zhao, X. B.; Zhang, L. H.; Liu, D. H. *Bioresour. Technol.* 2008b, 99, 3729-3736.
Zheng, Y. Z.; Lin, H. M.; Wen, J. Q.; Cao, N. J.; Yu, X. Z.; Tsao, G. T. *Biotechnol. Lett.* 1995, 17, 845-850.
Zheng, Y. Z.; Lin, H. M.; Tsao, G. T. *Biotechnol. Progr*. 1998, 14, 890-896.
Zheng, Y.; Zhang, R. H.; Pan, Z.; Labavitch, J.; Wang, D.; Teter, S.; Jenkins, B. *Appli. Biochem. Biotechnol*. 2007, 136/140, 423-435.
Zheng, Y.; Delwiche, M. 2008, Personal communication at UCDavis.
Zheng, Y.; Pan, Z.; Zhang, R.H.; Wang, D.; Jenkins, B. *Appli. Biochem. Biotechnol*. 2008, 146, 231-248.
Zhu, Y. M.; Lee, Y. Y.; Elander, R. T. *Appl. Biochem. Biotechnol*. 2004, 117, 103-114.
Zhu, Y. M.; Lee, Y. Y.; Elander, R. T. *Appl. Biochem. Biotechnol*. 2005a, 121, 1045-1054.
Zhu, S. D.; Yu, Z. N.; Wu, Y. X.; Zhang, X.; Li, H.; Gao, M. *Chem. Eng. Commun*. 2005b, 192, 1559-1566.
Zhu, S. D.; Wu, Y. X.; Yu, Z. N.; Liao, J. T.; Zhang, Y. *Process Biochem*. 2005c, 40, 3082-3086.
Zhu, L.; Dwyer, J. P. O.; Chang, V. S.; Granda, C. B.; Holtzapple, M. T. *Bioresour. Technol*. 2008, 99, 3817-3828.
Zhu, S. D. J. *Chem. Technol. Biotechnol*. 2008, 83, 777-779.

In: Bioethanol: Production, Benefits and Economics
Editor: Jason B. Erbaum
ISBN 978-1-60741-697-5

Chapter 2

MEETING EU BIOETHANOL TARGETS THROUGH IMPROVEMENTS IN WHEAT GRAIN QUALITY AND PRODUCTIVITY IN EUROPE

Richard M. Weightman and Daniel R. Kindred
ADAS, Centre for Sustainable Crop Management,
Battlegate Rd, Boxworth, Cambridgeshire, CB23 4NN, UK

ABSTRACT

Wheat is the most widely grown cereal worldwide (sown on over 200 million hectares), and can be grown throughout Eastern and Western Europe. Within Europe, winter wheat is an important feedstock for bioethanol production. However, concerns over land use change associated with biofuel crops, and of competition between food and fuel have raised questions over the desirability of producing ethanol within Europe. In the UK, average wheat yields have doubled from 4 to *ca.* 8 t/ha on farm since the 1960s, and individual farmers in the most productive areas can achieve yields of >10 t/ha. Given that the main driver for biofuels within Europe is reduction in greenhouse gas (GHG) emissions, high yields are an important element in delivering environmentally sustainable bioethanol. For example increasing yields from a 7.5 to a 10 t/ha wheat crop (with no change in inputs or grain composition) would lead to a positive effect on net GHG savings, with the percentage reduction in emissions (from bioethanol relative to petrol) increasing from 48% to 62%. FAO statistics suggest that wheat yields in Eastern Europe have fallen behind those in Western Europe since the late 1980s, from *ca.* 70% of UK wheat yields at their peak, to only 40% of UK wheat yields in 2007. Therefore there is large potential for improvement in wheat yields in a number of EU countries, as well as Ukraine and CIS. Bioethanol manufacture also yields a protein rich co-product (distillers dried grains and solubles; DDGS) which can be used to replace soya imported into the EU. FAO suggest that up to 13 million hectares of former arable land could be brought back into production within Eastern Europe. If this land were to be used for biofuel production, it would avoid the need for land use change outside Europe as the DDGS produced would substitute for imported soya. The available data therefore suggest huge potential through breeding and better agronomy of wheat to supply the demands of the bioethanol industry, while having minimal impact on food supply within Europe. However, significant barriers to increasing productivity still remain, in terms of higher input (fertiliser, fuel) prices and encouraging uptake of improved wheat varieties by the industry. A balance will need to be struck between the need for higher grain prices which will stimulate investment by agribusiness, and the low grain prices needed to make wheat bioethanol production competitive with that from sugar cane.

INTRODUCTION

With a population of almost 500 million in the 27 member states, the European Union (EU27) in 2008 had an average gasoline consumption of 5.7 litres per head of population per week, and a total consumption of 148 billion litres (bnL) per annum (Table 1.). Within the EU, two countries, the UK and Germany stand out clearly with the highest consumption figures of 8.2 and 7.4 L/person/week respectively. The UK and Germany together also account for 28% of the total population of the EU27. While petroleum and gasoline usage for transport in Europe is clearly large, it should be noted however that such numbers are dwarfed by the USA consumption of 43.7 L/person/wk, where there is much greater reliance on gasoline, compared to the greater use of diesel in Europe both for freight and private travel, and greater use of public transport. The main driver for bioethanol production within Europe is to reduce greenhouse gas (GHG) emissions, which has led to the setting of a target or 5% inclusion of biofuels in the road transport mix. Based on an annual motor gasoline consumption of 121 million tonnes (Mt) or 148 bnL in the member states of the EU27, the 5% target indicates an annual requirement of 6 Mt (7.7 bnL) of bioethanol, based on 2008 figures (Table 1).

Currently the actual supply of fuel ethanol against this 7.7 bnL target manufactured from indigenous crops in Europe is low (1.8 bnL/yr in 2008; www.ebio.org) the remainder being made up of imports - principally sugar cane bioethanol from Brazil. Fuel ethanol production accounts for only 1.6% of cereal grain production in the EU. The remainder of EU-produced grains supply traditional markets (*ca.* 90% for animal food, human food and industrial use). Therefore it is clear that for Europe as a whole, production will need to be expanded significantly to meet a 5% bioethanol requirement, particularly if this is to be supplied by cereals. While there are a number of economic reasons why bioethanol production in Europe is low, these are not considered here: The purpose of this paper is to simply address the question of whether it is possible to increase production within Europe, and using wheat as the principal feedstock, what barriers might need to be overcome.

Within Europe the UK is an interesting case study, because as noted earlier its population and gasoline consumption are both large and therefore demand for bioethanol is high. It also has the highest yields of wheat in Europe at 7.34 t/ha in 2007 (Table 2). For the UK, consumption of motor gasoline in 2007 was 20 Mt (Table 1) and a demand of approximately 3.5 Mt of wheat per annum would be created to meet the requirements of the Renewable Transport Fuel Obligation (RTFO). The UK currently has an export surplus of approximately 2 Mt of wheat per annum (Anon., 2008a) which could make up much of the requirement, with the balance being imported (most of the wheat bioethanol plants in the UK are planned for locations close to deep water ports). Alternatively the extra grain could be provided through increased production of home-grown wheat on previously set-aside land, or via replacement of other crops in the rotation.

Table 1. Petrol and gasoline consumption figures in 2007 for selected European countries and the USA, with the amount of ethanol required to meet the 5% inclusion level in the EU27

Region/Country	Petroleum consumption (Mt/year)*	Motor gasoline* (Mt/yr)	Motor gasoline* (ML/yr)	Population¶	Avg. gasoline consumption per head (L/pop/wk)	Ethanol req. at 5% inclusion (Mt/yr)
European Union (key wheat producing countries)						
Bulgaria	6	1	855	7,450,000	2.21	0.04
Estonia	1	0	407	1,330,000	5.89	0.02
Czech Republic	10	2	2,762	10,200,000	5.21	0.11
France	93	12	14,559	60,600,000	4.62	0.60
Germany	117	26	31,496	82,420,000	7.35	1.29
Hungary	8	2	1,996	10,000,000	3.84	0.08
Poland	25	4	5,387	39,000,000	2.66	0.22
Romania	10	2	2,170	22,300,000	1.87	0.09
United Kingdom	83	21	25,274	59,500,000	8.17	1.04
EU27	701	121	148,006	499,700,000	5.70	6.07
Europe (key non-EU wheat producers)						
Belarus	8	1	1,628	10,300,000	3.04	
Ukraine	17	5	6,027	47,400,000	2.45	
Russia	134	29	35,187	143,000,000	4.73	
Non European						
United States	985	436	531,809	295,000,000	34.67	

*, Petroleum and gasoline consumption statistics from EIA, Dec 2008 (www.eia.doe.gov)
¶, 2005 Population data from MacKay (2009).

Table 2. Current level of wheat production in selected European countries and quantity required to supply ethanol at 5% inclusion level

	Wheat production in 2007*				
Region/Country	Area (Mha)	Yield (t/ha)	Quantity (Mt)	Grain required at 5% inclusion level of ethanol (Mt/yr)	Quantity of grain for ethanol as prop of total wheat production (%)
Bulgaria	1.09	2.20	2.39	0.12	5.01
Czech Republic	0.81	4.88	3.95	0.39	9.80
Estonia	0.09	3.43	0.31	0.06	18.49
France	5.32	6.25	33.25	2.04	6.14
Germany	3.00	7.11	21.33	4.42	20.70
Hungary	1.11	3.59	3.98	0.28	7.02
Poland	2.13	3.94	8.39	0.76	9.00
Romania	1.90	1.50	2.85	0.30	10.67
United Kingdom	1.82	7.34	13.36	3.54	26.53
EU27	24.81	4.55	112.87	20.75	18.39
Belarus	0.43	3.28	1.41		
Russia	24.45	2.02	49.39		
Ukraine	6.02	2.29	13.79		

* , Source: FAO statistics (www.faostat.org).

However, while the UK can be partially self sufficient at current levels of production, these volumes of grain imply that current tonnages of wheat grain currently exported from the UK will certainly have to be satisfied by extra grain production elsewhere in the world. Again the volumes required are large: 3.5 Mt of wheat grain accounts for up to 27% of UK production and would have required 0.5 Mha of land in 2007 (Table 2). There is clearly a case to be made for increasing wheat productivity both within the UK and in Europe if we are to provide bioethanol from indigenous crops, and avoid both significant land use change and competition with food crops.

INCREASING PRODUCTION

Increasing production does not simply mean increasing the area of wheat sown. We have historically seen in the UK, that more cereals have been grown on less land (Figure 1) and this has been achieved through increasing grain yield per unit area.

Average UK wheat yields doubled from *ca.* 4 t/ha in the 1960s, to the current average of around 8 t/ha (Figure 2). Such yield improvements have been achieved through application of higher inputs, better agronomy and introduction of improved varieties, and this trend could theoretically continue. Alternatively, the concentrations of starch and sugars within the wheat grain could be increased for the same grain yield potential, to boost the alcohol yield (AY) per tonne of grain, thereby increasing the production of alcohol per hectare (alcohol productivity, AP).

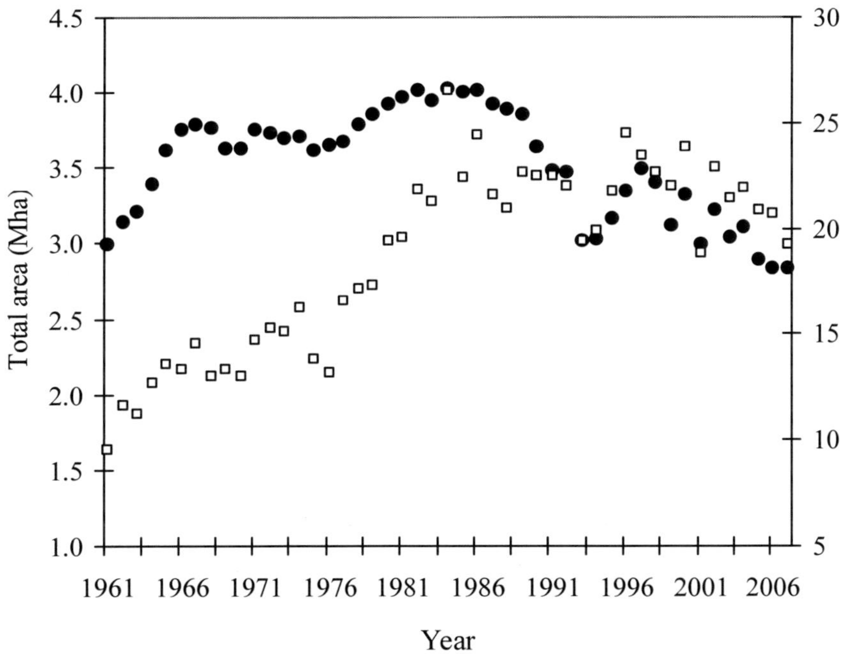

Figure 1. UK harvested area (●) and production of cereals (□) with time from 1961-2007 (Source: FAO Stats).

Most likely over time, improvements in productivity will come through a mixture of both genetic and agronomic approaches (focussing on both yield and grain composition).

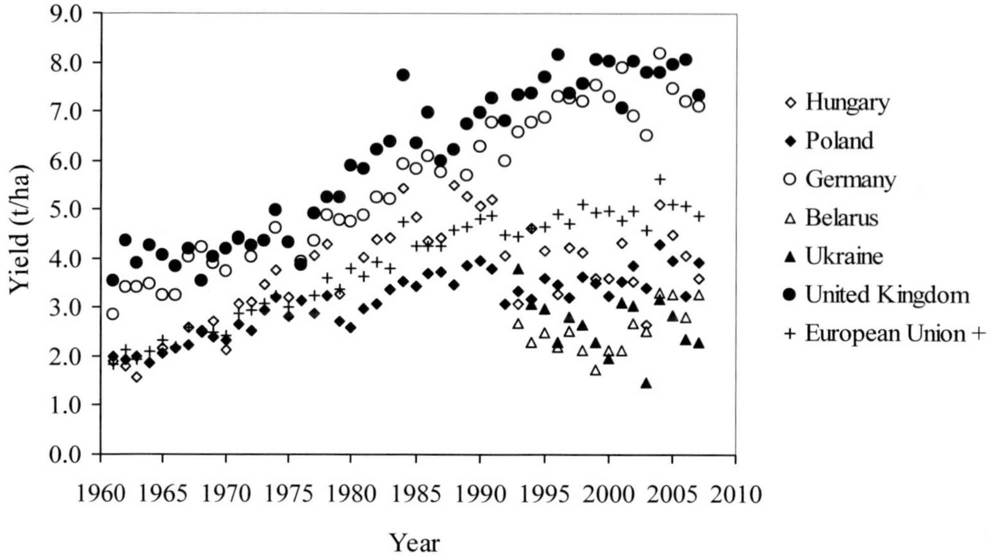

Figure 2. Wheat grains yields from selected Western and Eastern European countries from 1961-2007 (Source: FAO Stats).

Table 3. Grain yield, alcohol yield, and alcohol productivity of current benchmark wheat in the UK, and a theoretical wheat crop designed to match high alcohol productivity (High AP) of sugar cane

Trait	Current wheat (2007 average yields)			High AP wheat
	Ukraine	*Europe: EU27*	*UK*	
Starch content (%, dry basis)	69	69	69	80
Sugar content (%, dry basis)	3	3	3	3.5
Losses during conversion and fermentation (%)	14.7	14.7	14.7	8.0
Ethanol mass/grain mass (kg/t, dry basis)	344	344	344	430
Ethanol volume/grain mass (L/t, dry basis)	435	435	435	545
Ethanol/starch (L/10 kg)	6.0	6.0	6.0	6.5
Wheat grain yield (t/ha, 15%MC)	2.29	4.55	7.34	12.0
Alcohol productivity (L/ha)	847	1682	2714	5555

For the purposes of this analysis, it is useful to start with the question: What is a realistic target for AP of wheat ? If we were to match the most productive current biofuel crop (sugar cane) with wheat, an AP of around 5,500 L/ha would be a plausible, if stretching target biologically. Currently the AP of UK wheat is only 2,714 L/ha at the farm yield level, taking average farm yields (FAO stats) and an AY of 435 L/t of dry grain (370 L/t fresh basis; Smith *et al.*, 2006). Alcohol productivity of UK wheat is shown in Table 3, along with the impacts of grain yield on AP for 'very low' yielding (e.g. Ukrainian: 2.29 t/ha, 847 L/ha) and 'low' yielding wheat (EU27 average: 4.55 t/ha, 1,682 L/ha), with the same grain composition and AY. For examples of grain yields of other key EU wheat producing countries, see Table 2. The impact of grain yield on AP is marked, which highlights the importance of yield improvement to boost bioethanol production. Increasing yield is important also to minimise the carbon intensity of bioethanol production per unit area and so maximise the GHG savings, and this is considered in more detail later. However, to reach the target for a 'high AP' wheat of 5,500 L/ha it would be necessary to both improve yields levels by *ca.* 4 t/ha (based on UK wheat values), and to increase the starch and sugar concentration in the grain by about 95 kg/t dry grain. The potential for both increasing yields and changing grain composition are considered below.

Potential to Increase Wheat Alcohol Production without Additional Changes in Grain Composition

If we assume that current cropping patterns will hold true in the short term (for rotational and/or climatic reasons) then only a proportion of the 6 Mt of bioethanol in Europe will of necessity come from wheat. The remaining proportion will come mainly from corn and sugar beet. Currently these three crops are distributed in the following proportions: 77.4, 16.8 and 5.8% respectively by area grown in the EU27 (Source: FAO statistics). If the average yields in each of the EU27 for these three crops in 2007 are taken and multiplied by their respective AY (Wheat, 370 L/t; Maize, 417 L/t; Sugar beet, 95.2 L/t: Anon 2008c) the potential amounts of alcohol (in L/country) can be used to calculate the proportion of alcohol which would be supplied in each country by these three crops. Because of the high AY of sugar beet, the impact of wheat is somewhat reduced when compared to its area, but the aggregate figures still suggest that the majority (66.1%) of the increased demand for ethanol across the EU27 would be supplied by wheat. This is an average figure made up of some countries which grow a large proportion of wheat (e.g. 100% in Estonia) and others which grow little wheat (e.g. Portugal, 34% by area as a proportion of the three key bioethanol crops). The proportion of potential bioethanol production which would be supplied by wheat in the major EU wheat growing countries is shown in Table 4, the remainder again being provided by maize and sugar beet (and ignoring crops for potable alcohol such as rye, barley or grapes and alternative feedstocks such as surplus wine). Application of the 66.1% average figure to the EU27 requirement brings the realistic supply of bioethanol from wheat down to 4 Mt (5.1 bnL) and the demand for wheat grain down to 13.7 Mt which is still significant: about 12% of annual EU wheat production based on 2007 figures.

Table 4. Amount of wheat grain likely to supply ethanol in Europe based on current proportion of wheat in the rotation in selected EU countries and current typical ethanol yields

Region/Country	Likely proportion of ethanol supplied from wheat (%)*	Wheat grain needed to supply 66% of requirement at the the 5% inclusion level (Mt/yr)	Likely proportion of wheat for ethanolas proportion of total wheat production (%)
Bulgaria	87.0	0.10	4.36
Czech Republic	74.5	0.29	7.30
Estonia	100.0	0.06	18.49
France	59.0	1.20	3.62
Germany	66.7	2.95	13.81
Hungary	28.5	0.08	2.00
Poland	64.1	0.48	5.77
Romania	39.8	0.12	4.25
United Kingdom	88.9	3.15	23.58
EU27	66.1	13.72	12.15
Belarus	47.5		
Russia	80.5		
Ukraine	53.6		

*, Based on respective harvested areas and 2007 yields in each country for major ethanol feedstocks wheat, maize, sugar beet (source: FAO statistics) and ethanol yields on a fresh weight basis for the individual feedstocks: Wheat, 370 L/t; Maize, 417 L/t; Sugar beet, 95.2 L/t (source: UK Renewable Fuels Agency, Anon 2008c).

In the case of the UK alone, the comparable figures would be 88.9% of the supply of ethanol coming from wheat, giving a demand of 0.9 Mt of bioethanol, which would be supplied by 3.2 Mt of wheat from 440,000 ha of land, or approximately 24% of UK wheat production in 2007.

Based on the existing 24.8 Mha of wheat grown in the EU27, an increase in grain yield of around 12% will be needed to supplement the existing 113 Mt of wheat grown. Alternatively, based on an average annual yield of 4.55 t/ha in the EU27 (Table 2) an additional area equivalent to 3 Mha of wheat will need to be planted within Europe to meet the additional requirement assuming no grain yield increases. This section discusses the potential to achieve each of these options. The ability to increase grain yields needs to be considered separately for the higher yielding countries (UK, Germany, France) from those in Eastern Europe (e.g. Hungary, Poland, Belarus, Ukraine) where the average yields are nearer 2 t/ha.

INCREASING GRAIN YIELDS

Increasing Grain Yield – Western Europe

In the UK it has been noted that the annual average yield is 8 t/ha, and while this sounds impressive, it is much less than potential yield, given that the best farmers and also in experimental plot yields we achieve yields in the range 10-12 t/ha, close to that required by 'high AP' wheat (see Table 3). Moreover, Sylvester-Bradley *et al.* (2005) proposed that the theoretical potential for wheat in the UK is 19.2 t/ha. We need to understand therefore why actual yields appear so low. If, instead of apparently plateauing, yields had continued to increase in the last 10 years at the same rate they did between 1975 and 1995 (+0.15 t/ha/annum; Figure 1) then by 2007 we should have been seeing average farm yields of around 9.6 t/ha. Given that average farm yields are made of up of an aggregate of wheats grown in different positions in the rotation, two thirds of the wheat crops might be second wheats, which in broad terms have a yield potential *ca.* 1 t/ha less than first wheats. Therefore the average farm yield for first wheats should be in the region of 10-10.5 t/ha, which would put the best growers, particularly those in the main cereals growing regions in the Eastern region within striking distance of the 12 t/ha target (see Table 3).

Clearly in the mid-1990s something changed which caused yields to stop increasing at the rate they had previously. The reasons for the lack of yield improvement in the past 10 years is currently the subject of debate amongst crop scientists. From a technical viewpoint a number of contributory factors can be identified including: Declining genetic gains due to insufficient public investment in basic research underpinning commercial plant breeding; Poor profitability of commercial plant breeding (Caligari *et al.*, 2002) and conflicts for resource use between improving traits for yield, pest/disease resistance and environmental targets; Reductions in nutrient inputs (nitrogen use had plateaued by the end of the 1980's and sulphur deposition stopped as power station flue gases were cleaned up); Reduction in public support for an agricultural extension service (the services in England and Wales were privatized in 1997); Constraints on applying new technologies (e.g. biotechnology).

In contrast, Shepherd (2008) makes the case that the main reason for the yield plateau in UK wheat was not for technical, but for economic reasons. He demonstrates that following the period from 1981 to 1995 when UK wheat prices were consistently above 110 £/t (occasionally reaching 130 £/t) wheat prices then dropped to below 70 £/t and remained in the range 65-90 £/t until 2007 as intervention support reduced and UK wheat prices were left to adjust to world prices (Figure 3). In addition to a reduction in intervention support, the McSharry reforms brought in set-aside which was introduced in the harvest year of 1993 at which time 15% of cropped farm land had to be set-aside and this had reduced to 10% by 2006 (www.ukagriculture.com). It might have been expected that a yield increase would have been seen in response to introduction of set-aside if the worst land had been taken out of production, but this was not apparently the case. It may be that the reforms worked in clearly sending a message to farmers that increases in production were no longer desirable. Despite the introduction of a 0% set aside requirement in 2008 in response to the world food shortage, more recent EU agricultural reforms decoupling farm payments from crop production (removal of area aid payments) means that the cycle of continuing improvements in crop productivity have been broken. Farmers will certainly have invested a proportion of their income in environmental schemes in recent years, for which they now receive support payments from the public purse.

Historically in the UK, wheat yields viewed over a longer period than the 20 year period above, increased on average by *ca.* 1 t/ha/decade (Sylvester-Bradley *et al.*, 2005). Even if it was possible to turn on investment in plant breeding and increase the rate of release of higher yielding varieties while developing new crop protection products, the challenge of a yield increase of 12% for the average farm crop by 2010, or even within the next five years, seems to be difficult without the necessary political will to support agricultural production and give the industry the confidence to invest. If it is difficult to scale up production within Western Europe where structural reforms have weaned farmers away from a mindset of maximising yields, then some if not all of the additional production must come from other producers, principally in Eastern Europe.

Increasing Grain Yields – Eastern Europe

In Western Europe, the majority of wheat sown is 'winter wheat' i.e. sown in the autumn where, with mild winters, the crops can establish early and achieve a long growing season, with its main growth period in the following spring and summer, the main phase of growth lasting approximately 100 days (Sylvester-Bradley *et al.*, 2008a). In Eastern Europe with a continental climate, more spring wheat is sown with a shorter growing season and hence lower yield potential. Based on UK yields, spring wheats would be expected to yield on average 25% less than a winter wheat (Anon., 2008a). Yields of winter wheat where grown in Eastern Europe may also frequently be limited by harsh winters which cause very low plant populations coming into spring. Despite the lower yield potential which could be attributed to spring wheat, some of the key wheat producers within Europe would appear to have significant scope to increase yields, simply based on their historic performances (Figure 2). It is clear that after the late 1980's, wheat yields collapsed. Despite the inclusion of many former soviet block countries within the expanding EU in recent years, yields have not returned to pre-1990 levels. When the average yields of the four example countries Hungary, Poland, Belarus and Ukraine are expressed as a percentage of the UK yields (Figure 4) it can be seen that at their peak in the late 1980s, their grain yields had reached about 70% of the UK values, and appeared to be catching up. However, since then average yields have fallen back to settle at *ca.* 40% of UK wheat yields.

At the end of the 1990s Dyson (1999) suggested that following the collapse of Communism, recovery was under way in most of Eastern Europe, and that eventual membership of the EU augured well for countries like Poland. Unfortunately, the East European countries which have joined the EU27 have shown little sign of recovery in agricultural output although Poland has stabilised, albeit from a lower starting point than Hungary (see Figures 3 and 4). The European Bank for Reconstruction and Development (ERBD) suggest that yields in Ukraine and Russia will increase by around 11% between 2008 and 2016 (Anon., 2008d), but this is only a modest level of increase (+1.4% per annum from a very low yield base) and insufficient to provide the extra 13.7 Mt needed by 2010. The reasons for this lack of recovery to date point to a reduction in investment in agriculture both in terms of inputs of fertilisers and agrochemicals, as well as capital investment in machinery and storage facilities.

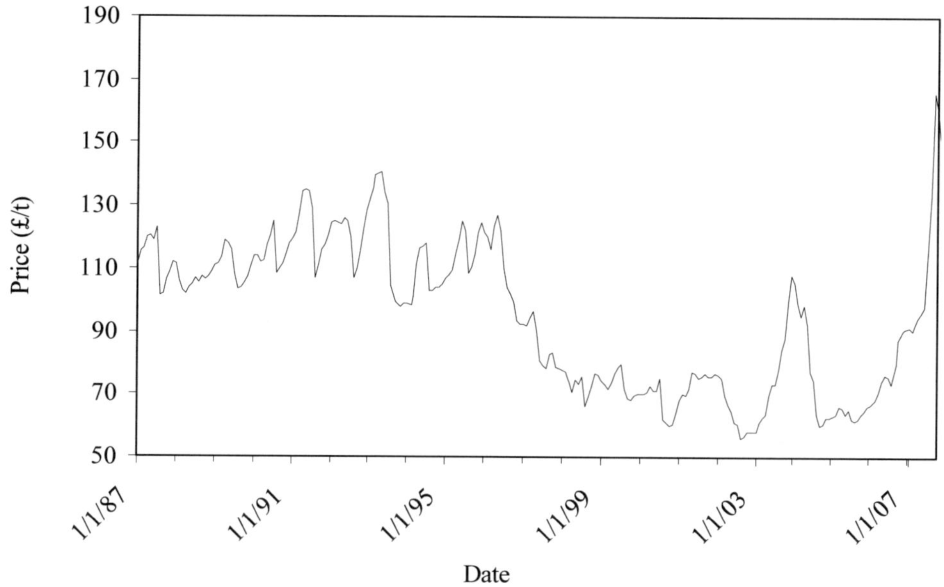

Figure 3. Historic feed wheat prices in the UK 1987 – 2007 (Source: HGCA).

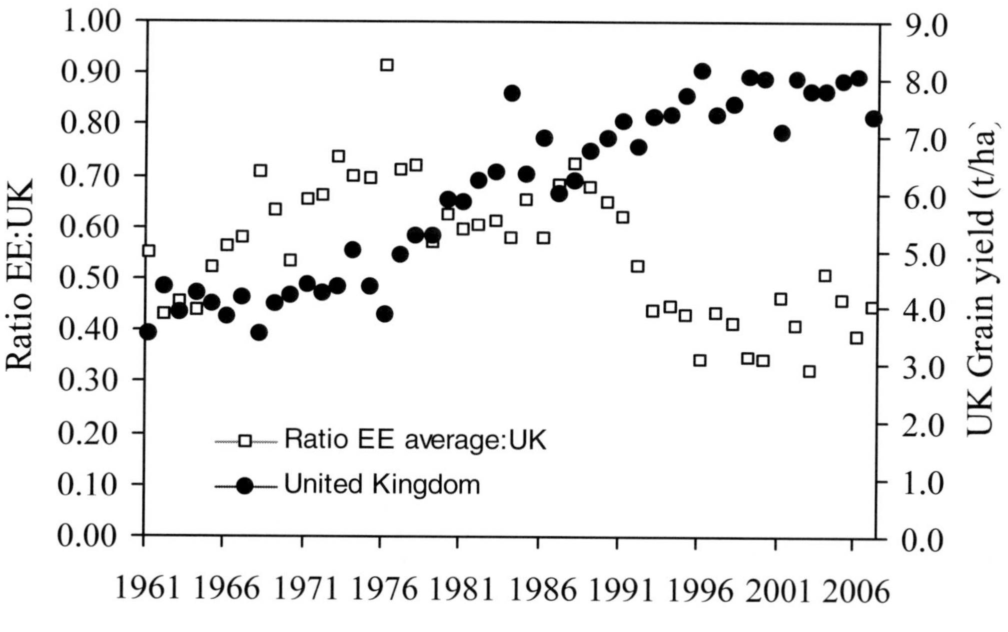

Figure 4. Ratio of average wheat grain yield for selected Eastern European (EE) countries to UK yields, plotted with UK wheat yields (EE countries selected: 1961-2007, Hungary, Poland; 1993-2007, Belarus, Ukraine; Source: FAO Stats).

Grzebisz and Fotyma (2007) stated that in the transition from planned to free market economies, the rise in fertiliser prices led to a decrease in fertiliser usage in the Central-Eastern Europe countries and that the decline in crop yields between 1988 and 2004 was strongly related to soil potassium (K) fertility status. The greatest drop in yield in the period 1990-1993 occurred 3 or 4 years after the cessation of high K use. If the transition to a free market economy elevated fertiliser prices in the early 1990s for the former communist block countries, then recent events in the commodity markets will only have exacerbated the problem. Nitrogen price closely follows energy prices and the cost of ammonium nitrate rose from €188 to 469 per tonne between February 2007 and August 2008. Similarly the price of potash increased from €175 to 725 per tonne, and triple super phosphate from €180 to 850 per tonne in the same period. There is potential to mine more potash – but the auction of licences for new potash mines in Russia has resulted in huge costs for investors there. It has also been stated by the fertiliser trade that phosphate supply is an increasing problem, and that phosphate supplies will run out within two generations. Undoubtedly yields could be increased in Eastern Europe, and there is anecdotal evidence that yields of wheat of 6 t/ha were achieved in specific areas of Ukraine in 2008 (K Dawson, pers. comm.) but the challenge is how average yields can be improved at a country level, and maintained over successive seasons.

Increasing Area of Land Sown to Wheat

Increasing Area in Western Europe

Releasing the 10% of arable land set-aside in 2006 should theoretically release *ca.* 2.5 Mha of land for wheat production based on the total area in the EU27 (Table 2). This is reassuringly close to the target of 3 Mha required, but there is no guarantee that it would yield as well as the most productive arable land. Some set-aside was based on headlands, and in many cases farmers may have put their worst land down to so-called ‘permanent’ set-aside. Moreover to bring an extra 3 Mha of arable land into production with the EU27 without causing land use change which may be unacceptable to the general public may be difficult. Wild places are valued and preserved, and moreover much of the land which was diverted to set-aside during the 1990s has reverted to semi-wild but managed habitat such as ‘conservation headlands’ which support biodiversity. Under CAP reform, farmers receive payments for management to meet environmental targets which supplement farm income. Farmers and land managers may be unlikely to want to return this land to arable production without a clear financial incentive, and where they do, at least 25% of the area would be sown to break crops, particularly oilseed rape in the case of the UK.

Increasing Area in Eastern Europe

An alternative to finding an additional 3 Mha of arable land in the EU27, would be to bring 6.8 Mha of land into production in former CIS countries (based on a lower yield of *ca.* 2 t/ha in 2007; wheat yields for Belarus, Russia and Ukraine are shown in Table 2). According to EBRD (Anon., 2008d) there are 13 Mha of idle arable land which could easily be brought back into production given the right investment. This appears to be a good fit with the required need for land for bioethanol. This land is reported to be 'non-marginal', and actually only half of the 26 Mha of land in Russia, Ukraine and Kazakhstan which was lost from production during the transition to free market economies.

It should be noted that many of the barriers discussed above in relation to expanding yield, apply also when trying to expand the area of land sown to wheat. For instance opening up new areas of arable land requires inputs of fertiliser and possibly liming materials. Therefore, these costs will be appreciable in an era of rising fertiliser prices as highlighted above. EBRD (Anon., 2008d) also suggest that there has been a rise in oilseed production at the expense of cereals in CIS countries by 6.3 Mha over the past 15 years, as oil crops have found a ready export market to Europe, and have given better gross margins than cereals. It should be remembered that Europe also has a large demand for biodiesel, therefore while biofuel targets remain, expansion of oil crops will continue to compete with bioethanol crops for the available arable land.

Despite these uncertainties, it would seem that there is great potential to increase the supply of wheat to Europe for bioethanol from countries like Ukraine and Russia. However, without any parallel yield improvements this begs another question. Is East European wheat a good feedstock from a GHG savings point of view ? This is considered in the final section.

The above discussion shows that it is feasible to increase the area of land sown to wheat in Europe if there is an economic incentive for farmers to increase agronomic inputs and investment in the supply chain, particularly in Eastern Europe. However, changes to grain composition, while slower than increase in land area and possible yield improvements, will in the long term be an important component of providing a sustainable feedstock for alcohol production. This is not necessarily through the obvious effect of increasing AY per tonne of grain, but through the parallel reduction in nitrogen use that would be expected to occur by reducing grain protein content. This is discussed in the following section.

Improvement of Wheat Grain as a Feedstock for Bioethanol

While it is not our aim to discuss in detail the composition of wheat and its conversion to bioethanol, it is first necessary to examine the key factors associated with the wheat grain which address its suitability for bioethanol production. This paper specifically relates to common or bread wheat (*Triticum aestivum* L.) which is marketed with both hard and soft endosperm textures. We have been in a fortunate position in the UK that there has been a body of knowledge built up on wheat for alcohol production through the existence of the Scotch whisky industry. The current UK distilling industry uses in excess of 750,000 metric tonnes of soft wheat per annum for grain whisky and for potable alcohol production. Wheat became the preferred feedstock for alcohol production when it became a cheaper feedstock than imported maize in the 1990s.

The Recommended List for wheat varieties published by the UK Home Grown Cereals Authority (HGCA; Anon., 2008) contains a recommendation for distilling use for certain winter wheat varieties, notably those with soft endosperm texture which are screened annually by the Scotch Whisky Research Institute (SWRI) in Edinburgh using their in-house laboratory methodology (Agu *et al.*, 2006). The historic SWRI data on alcohol yield from wheat was reviewed by Smith *et al.* (2006). According to the SWRI lab method, the alcohol yield (AY) of wheat is *ca.* 435 L/tonne grain (dry weight basis) but it is clear that there is a large variation in alcohol yields around this mean. Specifically, AY is inversely proportional to protein concentration, and positively related to starch content (Kindred *et al.*, 2008b).

Frustratingly there is little reliable data on starch concentration in wheat and despite its importance to alcohol production, starch appears to be rarely measured. This is partly due to the laborious nature and unreliability of the analytical methods, but also because there has been little interest in starch content except perhaps for animal feeding. Smith *et al.* (2006) suggested that a typical wheat grain composition (i.e. for wheat fertilized at the economic optimum nitrogen (N) fertiliser rate; Sylvester-Bradley and Kindred (2008) with a grain protein content of 11.5%) would be expected to have a starch content of 69%, plus 3% free sugars available for conversion to alcohol. These values were collated from various sources, as no-one had measured all the components of wheat grain and reported them, accounting for 100% of the dry matter, although they broadly agree with figures presented recently by Hazzledine (2008). For comparison, McAloon *et al.* (2000) carried out a similar exercise for corn in the US, and suggested comparable starch and sugar contents of 70 and 2% respectively on a dry basis, for corn with lower protein levels (9.5%) the balance being principally made up from the higher oil content of corn.

Starch and Sugar Contents

The values for typical wheat grain composition presented by Smith *et al.* (2006) have not to date been challenged, although a recent reanalysis of earlier ADAS work (data from Moss *et al.*, 1999) suggests that wheat varieties at 11.5% protein show a range of starch contents; between 70 and 75%. Moreover there are a number of key genotypes including the variety Riband (formerly a good distilling wheat but now outclassed on the basis of yield) which exhibit starch contents up to 81% (Kindred *et al.*, 2008d) albeit with very low protein contents (<7%) indicating that these samples were probably under-fertilized and hence low yielding. However, the data do demonstrate the scope to reduce grain protein genetically, and increase starch content to levels required to produce our 'high AP' wheat (Table 3). The challenge for plant breeders is to redesign the wheat crop such that the plant no longer takes up an excess of nitrogen from the soil during growth, which it then sequesters as storage proteins in the grain, and to do so in a way as to maintain high yield. One possible way to achieve low protein high yielding grain, is to maintain high rates of photosynthesis during grain filling, through delaying canopy senescence. Fortunately there is variation for a 'stay green' trait in most cereals which historically has been linked with yield maintenance under drought, and Foulkes *et al.* (2007) have begun to identify Quantitative Trait Loci in winter wheat which may eventually be used in breeding programmes. Interestingly varieties like Soissons, a French wheat which tends to produce bold grains with high starch content, also tends to show a good flag leaf persistence, although it is not a preferred wheat for distilling as it has hard endosperm texture. Another way to reduce N in the plant would be to reduce the N content of the wheat stem, which appears to contain temporary stores of soluble carbohydrates at flowering, which are re-translocated to the ear during grain filling, but a large quantity of N which is not.

If starch is rarely measured and difficult to quantify, then free sugars are measured even less frequently. The 'high AP' wheat calls for a slight increase in sugar content, although realistically it does not matter if the last 0.5-1.0% comes from starch or free sugars. The non-starch polysaccharides (NSP) in the grain are another significant component of the grain (*ca.* 11% of dry matter) which currently do not contribute positively to bioethanol production. In our laboratory we have begun to quantify the levels of NSP and the arabinoxylans (which are the main components of wheat NSP) in UK wheat varieties (Weightman *et al.*, 2008). The arabinoxylans are structural (cell wall materials) rather than storage carbohydrates and are synthesised via a different biochemical pathway to starch. Moreover, the NSP are made of pentose rather than hexose sugars, and so remain unfermented by standard brewers yeast. There is clearly potential through manipulating grain composition to increase fermentable sugar content. Whether it is necessary through plant breeding to reduce NSP and increase starch, or in the long run to develop strains of yeast which can ferment pentose sugars is debatable. Nevertheless, it would seem that variation in NSP and arabinoxylan content exists in wheat varieties, and to a certain extent the Scottish distilling industry selects against this by preferring to source wheat varieties which tend to give lower residue viscosities (aqueous viscosity in wheat being strongly linked to presence of NSP; Weightman *et al.*, 2001).

Losses during Starch Conversion and Fermentation

The data in Table 3 indicate that the losses of carbohydrate during distilling are around 14.7% of the expected fermentable sugars. The industry view is that a loss of only 8% of fermentable sugars would needed to support yeast growth, and this lower figure is therefore the target in our 'high AP' wheat. However, the 'missing' 6.5% is significant and cannot at present be explained: It could be that the conversion of starch is not complete, and like native starch, residual starch is to our knowledge rarely measured or reported. Arvalis (Anon., 2006) reported data for DDGS from two factories in France varying between 3 and 11.7% in their starch content, clearly indicating less than complete conversion. Similarly, Hazzledine (2008) reports a range of 4.2 – 13.8% starch content in wheat DDGS. Kindred *et al.* (2008b) found a difference in the rate of conversion of starch to alcohol between Riband and Option (a hard wheat), being significantly higher in the latter.

This observation may imply that there was more starch damage in grinding the hard wheat in the lab assessment method for AY which made it more accessible to enzymes. However, until commercial wheat bioethanol plants are established, data on true conversion rates will be limited. Another source of loss might be that the yeast converts significant quantities of the sugars to non-ethanolic compounds such as organic acids and glycerol (Jaques *et al.*, 2003).

Finally there is a possibility that the type of protein and particularly the amount of soluble protein in the wheat grain may be an important factor in influencing conversion. This is because the yeast requires a soluble nitrogen source whereas the bulk of the wheat protein is present as fairly water-insoluble aggregates. In a commercial bioethanol plant it is assumed that any variation in nitrogen availability from the wheat would be eliminated through additions of protease and urea. However, it remains to be seen whether there is variation in the capacity of different wheat varieties to support yeast through supply of the right amounts and balance of amino acids.

Protein Content

Modern bread-making wheat has been selected under nitrogen-rich environments, mainly because traditional markets have required high protein grain for breadmaking, and because with the introduction of dwarfing genes and use of plant growth regulators ('straw shorteners' such as chlormequat) wheat has generally responded well in terms of grain yield to increasing rates of nitrogen (N) application. Therefore, even the feed wheats have been selected in an environment of high N inputs, without a signal from the market to reduce grain protein. However, for bioethanol production, the protein in the grain merely serves to dilute the starch. Moreover wheat protein does not have an ideal amino acid composition for animal feeding (the destination for the DDGS). Of the storage proteins in the grain, we have shown that the gliadins are quantitatively the most important, accounting for 40% of the grain storage protein (Kindred *et al.*, 2008d). Moreover, gliadins increase in parallel with fertiliser induced increases in grain protein concentration. It therefore seems that a useful target to reduce the N demand of the crop, and also to produce a lower protein grain better suited to bioethanol production and with a better amino acid composition in the DDGS, would be achieved through reducing the levels of gliadins. Through a joint industry funded and government sponsored project called GREEN GRAIN (2005-2009; see www.greengrain.org) we have begun to investigate variation in grain storage protein composition in a range of European wheat cultivars and mapping populations, as well as targeting the canopy traits to reduce N demand and excess storage during vegetative growth.

Fertilised at the economic optimum N fertiliser rate, UK wheat typically has a grain N concentration of ~2% (Sylvester-Bradley *et al.*, 2008b). This is low compared to wheat samples for the milling industry, where grain N 2.3% (13% protein based on N x 5.7) are desired and for which there is a premium. Conversely 2% is high compared to spring barley varieties with average grain N contents of *ca.*1.5% (range 1.45-1.58%; Anon., 2008a). Part of the reason for the lower N content in barley could be due to effective dilution of storage components by the husk formed from the fused lemma and palea surrounding the grain: The husk accounts for 6.8% of total grain weight (Price and Parsons, 1979, cited by Morrison, 1993), therefore for a 'naked' barley grain, the N content would be higher at *ca.* 1.62%. However this is still appreciably less than the typical N content of UK wheat as reported in the Recommended Lists (a difference of 0.4% N between wheat and barley meaning around 2.3% protein using a 5.7 conversion factor from N to protein).

A study of varietal effects on nitrogen requirements and nitrogen use efficiency (NUE) in wheat and spring barley has recently been carried out (Sylvester-Bradley and Kindred, 2009). It appears that breeding has increased grain yields of both wheat and barley, and so NUE at a given N rate has increased. However, whilst N requirements of wheat have increased, those of barley have remained constant. The NUE at the optima in wheat has therefore not improved, whilst it has improved significantly in barley. The improved NUE in barley is associated with lower grain N%, and may be due to differences in the breeding strategies between wheat and barley: wheat is selected for high grain N for breadmaking, barley is selected for low grain N for malting; wheat is selected when grown with high levels of N fertiliser, barley is selected under lower N regimes; barley breeding tends to use one programme for malting and feed varieties, wheat breeding crosses and early selections tend to be focussed on breadmaking.

In contrast to the UK, in Denmark where N fertiliser use has been constrained for several years for environmental reasons, there are now wheat varieties on the Danish recommended list with grain N contents as low as 1.63%, yet farm yields still averaged 7.1 t/ha over the ten years between 1998 and 2007 (FAO Stats). Therefore it is clearly possible today to generate moderately low protein grain with respectable grain yields. In Denmark, the variety evaluation system favours wheat with a low N requirement, and the country produces very good yields compared to the EU27 average. To our knowledge Denmark is the only country which has tried to reduce N inputs into agriculture in this way, and such a system would be expected to produce a very good feedstock for wheat-bioethanol production. Unfortunately, in Denmark the vast majority of the wheat is used for feeding to pigs, and there are no plans to use wheat for bioethanol production in that country (Bonde Peterson, pers comm.).

With the exception of Denmark, currently there are no financial incentives or legislative drivers for growers or wheat breeders to produce low protein wheat in any European member state. Therefore farmers tend to aim for high yields by fertilising at the recommended level in their particular country, and aiming to over-fertilise in many cases, with the hope of producing high protein grain which might capture a premium in the bread market. In itself, this is not a problem if the bioethanol producer does not wish to pay a premium for low protein/high starch grain, because as we have seen, the main driver of alcohol productivity per unit area is grain yield. However, where it causes a problem is with the assessment of the benefits of bioethanol production in terms of GHG savings. Here N fertilization is critical and this is discussed in the next section.

Opportunities for Greenhouse Gas Savings from Wheat Bioethanol

As mentioned earlier, the main reason there is a market for bioethanol in Europe is to minimise GHG emissions associated with transport fuel use. Therefore the GHG savings associated with bioethanol production needed to be considered carefully. Savings for bioethanol from UK wheat can range from 10–95% depending on processing options and grain production (Woods *et al.*, 2008) with crop production typically accounting for 50–80% of the GHG emissions from the biofuel. The vast majority of emissions from crop production are associated with N fertiliser, both from its manufacture and from related soil emissions of nitrous oxide (N_2O) (Kindred *et al.*, 2008a). The very high emission factors from both manufacture and N_2O emissions means N fertiliser makes up around 70% of the GHG costs of growing a tonne of wheat. It should be clear therefore that reduction of N inputs during crop production, as well as being a route to producing a better (high starch) feedstock for the bioethanol producer, will also give a feedstock with greater potential benefits in terms of GHG emissions, at a given level of yield.

NITROGEN FERTILISER EFFECTS ON WHEAT

Nitrogen Use and Grain Yields in UK Wheat

Recommended fertiliser N rates in the UK are determined from measured grain yield to N response curves, and consideration of the breakeven ratio between grain and nitrogen i.e. the weight of grain which is required to pay for a unit of N fertiliser. For many years, the breakeven ratio used in the UK fertiliser recommendations was 3:1 (Anon., 2000). As N prices have increased in recent years, the breakeven ratio has increased to be closer to 9:1 and all other things being equal, this results in the economic N rate coming down i.e. the grower is advised to put on less fertiliser N. This may be beneficial for the bioethanol producer in terms of producing lower protein grain, but the grain becomes more expensive to purchase because the grower has to be compensated for the lower yield. In other words, the bioethanol producer might have to pay a financial premium for low protein grain (Sylvester-Bradley and Kindred, 2008).

Table 5 considers the wider implications of N fertiliser use on GHG savings and land use for bioethanol production from UK wheat. The data are based on yield levels which correspond to typical UK performance, and are based on the N response data of Kindred *et al.* (2008b). The yields in Table 5 broadly agree with the national average for 2007 of 7.34 t/ha (assuming an economic optimum N rate applied). As the rate of N fertiliser is increased, the wheat shows an increase in grain yield from 5.17 t/ha with zero N applied up to a peak of 7.91 t/ha at a N fertiliser rate of 160 kg/ha, and then a fall to 7.37 t/ha in the highest N treatment (240 kg/ha). The yield decrease at high N rates is typically caused by lodging when crops are over-fertilized. It can also be seen that as N rate increases, so does grain protein concentration (from 7.45 to 11.80%) while AY decreases from 466 to 435 L/t. Because of the large effect of grain yield on alcohol productivity (and despite the decrease in AY per tonne) the AP increases to a peak of 3,006 L/ha at a fertiliser N rate of 160 kg/ha (close to the economic optimum for grain yield) and then decreases in the same way as grain yield, at the highest N rates.

Using these basic data it was thus possible to estimate the GHG savings associated with bioethanol production using the UK Renewable Fuels Agency (RFA) Carbon Calculator (www.rfa.gov.uk; Anon, 2008d). The savings in emissions reduce steadily from 72% with zero N applied, to 16.5% at the highest N rate applied, the GHG savings clearly being dominated by the emissions associated with N fertiliser use (Table 5). Simplistically it would therefore appear that the best biofuel would be produced with zero N applied. However, at this level, the lowest grain yields and AP are also realised and therefore it would require more land to satisfy the required demand for ethanol. It can be seen that 47% (183,000 ha) more land would be needed at zero N applied to be able to match the AP of 3,006 L/ha produced at the maximum point (at N rate 160 kg/ha). Clearly using very low N inputs requires more land to grow wheat for bioethanol, and so in the UK the extreme situation of applying zero N (as might be the case in organic production) using these data would require 577,000 ha, rather than the 394,000 ha of land needed near the current economic optimum N rate. There is another benefit of growing wheat for bioethanol, and that is to produce the DDGS, a protein-rich source of animal feed which can be used in Europe to displace imported protein such as soya.

Table 5. Typical net GHG savings with various yield scenarios and N inputs for wheat variety Option grown at ADAS Rosemaund in Herefordshire UK with an optimum economic optimum N rate of 153 kg/ha

Outputs, GHG savings and land use requirements	N fertiliser (kg/ha)						
	0	40	80	120	160	200	240
Outputs							
Grain yield (t/ha @85%DM) at farm level*	5.17	6.36	7.13	7.63	7.91	7.69	7.37
Grain protein (%DM)	7.45	8.25	9.08	9.82	10.13	11.34	11.80
AY (L/t DM basis)	466	461	455	449	447	438	435
AP (L/ha)	2051	2490	2754	2915	3006	2867	2727
AP (t/ha)	1620	1967	2175	2303	2375	2265	2154
DDGS per unit wheat feedstock (t/t grain)	0.374	0.381	0.389	0.397	0.400	0.411	0.416
DDGS per unit ethanol (t/t EtOH)	1.19	1.23	1.28	1.31	1.33	1.40	1.42
DDGS per unit area (t/ha)	1.93	2.43	2.78	3.03	3.16	3.16	3.07
GHG emissions/savings¶							
Estimate of savings (% based on gasoline at 85 kg/GJ)	72.1	63.8	56.6	49.1	41.1	29.9	16.7
Total emissions per unit biofuel (kgCO_2e/t EtOH) exc. CPC	1218	1427	1613	1802	1989	2282	2594
Co-product credit (CO_2e/t EtOH)†	-582	-602	-623	-642	-649	-682	-694
Net emissions per unit biofuel (kgCO_2e/t EtOH) inc. CPC	636	825	990	1160	1340	1600	1900
Total emissions per unit energy from bioethanol (kgCO_2e/GJ) inc. CPC	23.7	30.8	36.9	43.3	50.1	59.6	70.8
Total emissions per unit energy from bioethanol (kgCO_2e/GJ) exc. CPC	45.4	53.3	60.1	67.2	74.4	85.0	96.7
Estimate of savings (% based on petrol at 85 kg/GJ) exc. CPC	46.6	37.3	29.3	20.9	12.5	0.0	-13.7
Implications for UK land requirements of 0.9 Mt ethanol from wheat							
Total land required at predicted level of AP ('000 ha)	577	475	430	406	394	413	434
Additional land required ('000 ha) in UK on top of minimum 394,000 ha	183	81	36	12	0	19	40
Proportion of current UK wheat area (%)	31.7	26.1	23.6	22.3	21.6	22.7	23.8
Quantity of DDGS produced (Mt)	1.11	1.15	1.19	1.23	1.24	1.31	1.33

*, Farm yields estimated from plot yields x 0.8 based on original data in Kindred *et al.* (2008b),

¶, GHG costs and savings calculated using RFA Carbon calculator (Anon., 2008c).

†, Co-product credit (CPC) estimated at -488 kg CO_2e/t DDGS (Anon., 2008c).

The DDGS is assumed therefore to reduce the land needed for crop production elsewhere in the world, and for this reason in the carbon calculator, the production of bioethanol can be assigned a co-product credit (CPC) of (-) 488 kgCO_2e/t DDGS (Anon., 2008d). Because grain protein and DDGS yield per hectare increase with increasing fertiliser N rate (Table 5) the co-product credit increases at the highest N rate which offsets some of the GHG emissions per tonne ethanol associated with higher N use. If the co-product credit is not taken into account, then at an N fertiliser rate of 200 kg/ha, the net GHG savings disappear altogether. Moreover, at an N rate of 240 kg/ha, the GHG savings are negative, or in other words, the GHG emissions associated with growing the crop are greater than the GHG savings from the bioethanol. One practical implication of this for Europe is that the DDGS produced also provides an opportunity to reduce indirect land use change (ILUC) by replacing imported soya. At the lowest levels of N applied, again less DDGS per hectare of land is produced (Table 5) which means there is less DDGS to replace imported protein, and so there would potentially be a greater impact of ILUC associated with production of protein crops outside Europe.

Analysis using a typical N response curve in the UK (Sylvester-Bradley and Kindred, 2008; Clarke *et al.*, 2008) showed that emissions are minimised with around 50 kg N/ha. However, this ignores the fact that less grain is produced per ha; if the world demand for grain is stable or increasing then the unrealised grain will be produced elsewhere as discussed above. Because LUC releases enormous amounts of CO_2 from soil and vegetation, the net effect can mean that it is better to use more N fertiliser to produce more yield and hence avoid ILUC. Optimising for GHG on this basis gives an N optimum around 150 kg/ha, but this is entirely dependent on the many assumptions made. It is possible that optima to minimise GHGs in wheat are actually higher than current economic optima, and no doubt further debate will continue on this important topic as the market and legislation underpinning it develops over the coming years.

Nitrogen Use and Grain Yields in East European Wheat

It was argued earlier that the quickest way to increase wheat production for bioethanol might be to use recently cropped, but currently unproductive land in Eastern Europe. However, consideration of the suitability of Ukrainian wheat for UK bioethanol production raises its own questions. The RFA carbon calculator has default values for Ukrainian wheat with the grain yield set at 2.6 t/ha (slightly higher than the FAO value; Table 2) and N inputs set at 90 kg/ha (approximately half that applied to a UK wheat crop) and phosphate and potash set at 80 kg/ha each. With this level of inputs and grain yield, the carbon intensity of the fuel supply chain is 2,760 kgCO_2e/t bioethanol, and there are no net benefits in terms of GHG emissions (GHG savings -21.7%). Part of the reason for high carbon intensity is the longer supply chain compared to UK wheat, including 1,700 km of transport of the feedstock over land and 2,700 km transport by sea which accounts for around 10% of the costs. However even removing all transport costs and manufacturing the bioethanol in Ukraine would still show no net benefit due to the very low grain yields. Using average grain yields for Ukraine of *ca.* 3.5 t/ha in 2008 (Chernysh pers. comm.) and the same levels of inputs would lead to small but positive net GHG savings of 2.8% while decreasing phosphate and potash inputs to 40 kg/ha increases the net GHG savings to 5% (including the transport costs).

Discussion with agronomists in the Ukraine suggest that on the more productive land, more realistic yields of *ca.* 6 t/ha can be achieved, albeit with higher N inputs. Readjustment of the RFA default values for Ukrainian wheat, using 6 t/ha grain yield and a nominal 130 kg/ha of fertiliser N would give a positive GHG saving of 20.5%. At present, there are few published, reliable data on wheat production in Ukraine. Clearly more work is needed to truly understand the capacity to expand wheat production in Eastern Europe, the range of yields which make up the average, and to understand the typical levels of inputs, particularly N.

Alternative Cereals to Wheat for Bioethanol in Europe

Consideration of the importance of N-use implies that we need cereals which yield well at low N applications. This was discussed earlier in the context of changing wheat composition, which through conventional breeding routes will of necessity take some time and is not a realistic solution to producing an optimal feedstock in terms of GHG savings by 2010. However there may be other shorter-term alternatives, by switching to other species. Triticale is a hybrid of rye and wheat which is popular in certain areas of Europe, but has never been favoured in the UK partly because of its lower yield potential (actually this only a 'perceived' lower yield potential, because triticale tends to be grown on the worst land, and so there are no fair comparisons with wheat).

Triticale is a soft grain without a husk, and therefore can replace wheat as a feedstock for bioethanol production, with comparable AY to wheat (Davis-Knight and Weightman, 2008). In terms of the costs of production, triticale is very good at scavenging soil N and the recommended levels of applied fertiliser N are much lower than wheat. For example, the highest rate of N fertiliser recommended for triticale in the UK is 130 kg/ha whereas for wheat the average is *ca.* 180 kg/ha, and the highest recommended rate 240 kg/ha (Anon., 2000). The implication of this is that at comparable yields, the net GHG savings associated with bioethanol from triticale would be greater than for wheat. Most interestingly, if we assumed that an increase in UK wheat output by 3.2 Mt/annum would in the short term come from an increase in area sown to wheat, this implies an increase in the frequency of wheat crops in the rotation. However, in a second cereal position (i.e. wheat after wheat), there is typically a yield loss of up to 1 t/ha due to build up of pests and diseases, principally 'take-all'. The loss in yield, and the fact that the grower may apply slightly more N fertiliser to a second wheat, means that the net GHG savings associated with bioethanol production are less than for a first wheat. However, triticale has much better take-all resistance than wheat, and together with its lower N inputs, the GHG savings associated with triticale could be much better in the second cereal position (32.9 *vs* 12.5% net savings from triticale and wheat respectively; Weightman and Davis-Knight, 2008). Notwithstanding the benefits in terms of reduced GHG emissions per tonne of ethanol produced, it should be noted that in this scenario, the triticale in the second cereal position might yield 1.2 t/ha less than a second wheat, with implications for the area required for bioethanol production, the amount of DDGS produced per unit area, and for ILUC issues.

Conclusion

The creation of a market for wheat for bioethanol is a positive challenge for the agricultural industry across Europe. However, as we have seen, the limitations on land area and the apparent plateauing of yields in Western Europe, together with a collapse in yields in Eastern Europe, mean that it is not wholly clear where the necessary increase in wheat production will come from to supply this new market. Increasing average grain yield at the national level may be one of the hardest challenges to overcome - not just producing high yield occasionally in specific locations, but producing it consistently across a rotation and large areas of a country.

What is clear is that there needs to be investment in research in Western Europe to enable farmers to increase grain yield while avoiding over-application of inputs, particularly of fertiliser N. In both Western and Eastern Europe, investment is required at a basic level to enable increasing inputs, and investment in machinery and storage that enable a more efficient and productive supply chain. Over time, farmers have shown themselves capable of rising to the challenge of raising productivity and changing cropping patterns to meet market requirements, but they can only invest when there are periods of high and stable prices.

The paradox of high grain prices seen in 2007-2008 is that while they are needed to stimulate investment by agribusiness, high feedstock prices can make bioethanol production from grains uneconomic. The paradox is much the same with grains for food – high prices stimulate investment and can provide export revenue for developing countries, while at the same time can have a negative impact on the poorer members of the community who spend a larger proportion of the income on food (Anon. 2008e).

Producing high yielding wheat in Europe will play an essential part in feeding a growing population, while at the same time allowing us to minimise GHG emissions associated with transport, and minimising imports of protein into Europe. Wheat has great potential as a sustainable feedstock for bioethanol in Europe assuming a favourable political and economic climate to support the investment needed over the coming years.

Acknowledgements

The authors would like to thank Oleh Chernysh of Kyiv Oblast Rural Advisory Service, and Peter Mitchell and Keith Dawson for information on wheat yields in Ukraine. The financial support of the UK Home Grown Cereals Authority during the underpinning research work cited in this paper is gratefully acknowledged.

References

Anon. 2000. *Fertiliser Recommendations*, 7th Edition. Ministry of Agriculture Fisheries and Food. London, The Stationary Office. ISBN 0-11-243058-9.

Anon. 2006. La letter d'information du Pole Valeurs Nutritionnelles d'ARVALIS-Institut du vegetal. No.11. November 2006. www.arvalisinstitutduvegetal.fr/fr/fichier/communique/332_Newslim_n%C2%B0_11.pdf.

Anon. 2008a. HGCA Recommended Lists 2008/09. London: HGCA www.hgca.com.

Anon. 2008b. Genstat 11th Edition. VSN International Ltd.: Hemel Hemstead, UK.

Anon. 2008c. Carbon and sustainability reporting within the renewable transport fuel obligation. Technical guidance part 2. Carbon reporting – Default values and fuel chains. Office of the Renewable Fuels Agency, v 1.3, August 2008, www.renewablefuelsagency.org.uk.

Anon. 2008d. Fighting food inflation through sustainable investment. Grain production and export potential in CIS countries. Rising food process: causes, consequences and policy responses. Report by European Bank for Reconstruction and Development and FAO. http://www.fao.org/newsroom/common/ecg/1000808/en/FAOERBD.pdf

Anon. 2008e. *The state of food and agriculture. Biofuels: prospects, risks and opportunities.* Food and Agriculture Organization of the United Nations, Rome. ISBN 978-92-5-105980-7. www.fao.org/catalog/inter-e.htm.

Agu R C, Bringhurst T A and Brosnan J M. 2006. Production of grain whisky and ethanol from wheat, maize and other cereals. *Journal of the Institute of Brewing 112*; 314-323.

Caligari P D S, Lobley K, Weightman R M, Sylvester-Bradley R and Temple M. 2002. *The role of future public investment in the genetic improvement of UK grown crops.* Final report on project ST0158 for the Department for Environment, Food and Rural Affairs. September 2002.

Clarke S, Kindred D, Weightman R M and Sylvester-Bradley R. 2008. *Growing Wheat for Alcohol and Bioethanol Production in the North East.* An ADAS report commissioned by One North East and North East Processing Industries Cluster. www.adas.co.uk/reports.

Cottrill B, Smith T C, Berry P, Weightman R M, Wiseman J, White G and Temple M. 2007. *Opportunities and implications of using the co-products from biofuel production as feeds for livestock.* HGCA Review 66. 95 pp. HGCA, London.

Davis-Knight H and Weightman R M. 2008. Triticale as a low input cereal for alcohol production. I. Alcohol yields and processing quality of triticale varieties grown under UK conditions. *Aspects of Applied Biology 90, Biomass and Energy Crops III*; 135-142.

Dyson T. 1999. Prospects for feeding the world. *British Medical Journal 319* (7125); 988-991.

Foulkes M J, Sylvester-Bradley R, Weightman R M and Snape J W. 2007. Identifying physiological traits associated with improved drought resistance in winter wheat. *Field Crops Research 103* (1); 11-24.

Gallagher E. 2008. *The Gallagher Review of the indirect effects of biofuels production.* Renewable Fuels Agency. London. www.renewablefuelsagency.org

Grzebisz W and Fotyma M. 2007. *Recommendations and use of potassium fertilisers in Central-Eastern Europe.* Proceedings of the Fertiliser Society, 621, 24pp. The International Fertiliser Society, York, UK. www.fertiliser-society.org.

Hazzledine M. 2008. *Nutritional and economic value of co-products from biofuel production.* Chapter 14. In Recent Advances in Animal Nutrition – 2008 (eds P C Garnsworthy and J Wiseman). Nottingham University Press, Nottingham, UK. ISBN 978-1-904761-044.

Jaques K A, Lyons T P and Kelsall, D R. 2003. *The Alcohol Textbook.* Nottingham, UK: Nottingham University Press.

Kindred D R, Smith T C, Sylvester-Bradley R, Ginsberg D and Dyer, C J. 2007. *Optimising nitrogen applications for wheat grown for the biofuels market.* HGCA Project Report No. 417. HGCA, London.

Kindred D R, Mortimer N, Sylvester-Bradley R, Brown G and Woods J. 2008a. *Understanding and managing uncertainties to improve biofuel GHG emissions calculations*. HGCA Project Report 435 Part II. HGCA, London.

Kindred D R, Verhoeven T, Weightman R, Swanston J S, Agu R C, Brosnan J and Sylvester Bradley R. 2008b. Effects of variety and fertiliser nitrogen on alcohol yield, grain yield, starch and protein content, and protein composition of winter wheat. *Journal of Cereal Science 48*, 46-57.

Kindred D R, Berry P, Burch O and Sylvester-Bradley R. 2008c. *Effects of nitrogen fertiliser use on green house gas emissions and land use change*. Aspects of Applied Biology 89, Effects of Climate Change on Plants: Implications for Agriculture. In press.

Kindred D R, Weightman R M, Clarke S, Agu R, Brosnan J and Sylvester-Bradley R. 2008d. Developing wheat for the biofuels market. *Aspects of Applied Biology 90, Biomass and Energy Crops III*; 143-152.

MacKay D J C. 2009. *Sustainable energy - without the hot air*. ISBN 978-0-9544529-3-3. UIT, Cambridge, England.

McAloon A, Taylor F, Yee W, Ibsen K and Wooley R. 2000. *Determining the cost of producing ethanol from corn starch and lignocellulosic feedstocks*. National Renewable Energy Laboratory report NREL/TP-580-28893. Golden, Colorado.

Morrison W R. 1993. *Barley lipids*. Chapter 5 in: Barley, Chemistry and Technology. Eds A W MacGregor and R S Bhatty, pp. 199-246. ISBN 0-913250-80-5. AACC Publications, St Paul, Minnesota, USA.

Moss A, Givens I, Froment M A and Lunn G. 1999. *Factors affecting the nutritive value of wheat for ruminants*. HGCA Project Report 182. HGCA, London.

Price P B and Parsons J G. 1979. Distribution of lipids in embryonic axis, bran-endosperm and hull fractions of hulless barley and hulless oat grain. *Journal of Agricultural and Food Chemistry 28*, 875-877.

Riddle J. 2008. *EBRD and FAO call for bold steps to contain soaring food prices*. http://www.fao.org/newsroom/en/news/2008/1000808/index.html

Searchinger T, Heimlich R, Houghton R A, Dong F, Elobeid A, Fabiosa J, Tokgoz S, Hayes D and Yu T-H. 2008. Use of U.S. Croplands for Biofuels Increases Greenhouse Gases Through Emissions from Land Use Change. Sciencexpress 10.1126/science.1151861.

Shepherd M. 2008. *UK crops for food and fuel*. A report prepared for NorthEast Biofuels, October 2008. NEB-FoodFuel-Report—final-final4.pdf available at: www.biofuels now.co.uk/filedownload.php?id=58.

Smith T C, Kindred D R, Brosnan, J M, Weightman R M, Shepherd M and Sylvester-Bradley R. 2006. *Wheat as a feedstock for alcohol production*. Research Review No. 61, HGCA, London.

Sylvester-Bradley R, Foulkes M J and Reynolds M. 2005. *Future wheat yields: Evidence, theory and conjecture. Chapter 11 in: Yields of Farmed Species – Constraints and opportunities in the 21st Century*. Eds R Sylvester-Bradley and J Wiseman. ISBN 1-904761-23-2. Nottingham University Press, Nottingham, UK. pp 233-260.

Sylvester-Bradley R. 2008. *Critique of Searchinger (2008) and related papers assessing indirect effects of biofuels on land-use change*. A report forming part of The Gallagher Biofuels Review: ADAS. www.renewablefuelsagency.org

Sylvester-Bradley R and Kindred D R. 2008. *Developing and growing wheat for the biofuels market*. In 'Arable cropping in a changing climate' HGCA R and D conference 23-24 January 2008, pp. 110-125.

Sylvester-Bradley R. and Kindred D R. 2009. Analysing nitrogen responses of cereals to prioritise routes to the improvement of nitrogen use efficiency. *Journal of Experimental Botany* (in press).

Sylvester-Bradley R, Berry P, Blake J, Kindred D R, Spink J, Bingham I, McVittie J and Foulkes M J. 2008a. *The Wheat Growth guide, Spring 2008 Second edition*. Agriculture and Horticulture Development Board, HGCA, London.

Sylvester-Bradley R, Kindred D R, Blake J, Dyer C and Sinclair A. 2008b. *Optimising fertiliser nitrogen for modern wheat and barley crops*. Project Report No. 438. HGCA, London

Weightman R M and Davis-Knight H. 2008. Triticale as a low input cereal for alcohol production. II. Potential to reduce greenhouse gas emissions relative to bioethanol from wheat. *Aspects of Applied Biology 90, Biomass and Energy Crops III*; 165-172.

Weightman R M, Forge C D and Quandalle C. 2001. A rapid viscometric tool for measuring feed wheat quality and the relationship between the quality of hybrid wheats and their parental lines. *Aspects of Applied Biology 64*; 79-84.

Weightman R M, Davis-Knight H, Wang R H, Misailidis N, Villanueva A L and Campbell G M. 2008. Towards defining optimal feedstocks for a wheat biorefinery: Co-production of arabinoxylans with bioethanol. *Aspects of Applied Biology 90, Biomass and Energy Crops III*; 153-160.

Woods J, Brown G, Gathorne-Hardy A, Sylvester Bradley R, Kindred D R and Mortimer N. 2008. *Facilitating carbon (GHG) accreditation schemes for biofuels: feedstock production*. Project Report 435. London: HGCA.

In: Bioethanol: Production, Benefits and Economics
Editor: Jason B. Erbaum
ISBN 978-1-60741-697-5

Chapter 3

DEVELOPMENT OF EXTRACTIVE PROCESSES AND ROBUST MATHEMATICAL MODEL FOR BIOETHANOL PRODUCTION

Rafael Ramos de Andrade[a], Elmer Ccopa Rivera[a], Daniel I. Pires Atala[b], Francisco Maugeri Filho[b], Rubens Maciel Filho[a] and Aline Carvalho da Costa[a]

aLaboratory of Optimization, Design and Advanced Control, School of Chemical Engineering, University of Campinas, 13081-970, Campinas, SP, Brazil
bLaboratory of Bioprocess Engineering, School of Food Engineering, University of Campinas, 13081-970, Campinas, SP, Brazil

ABSTRACT

Bioethanol production processes have been running for years and ethanol consumption has been proved to be advantageous over fossil fuels considering gas emission, especially in metropolitan areas, reduction of dependence of oil and gas imports and social aspects, such as job generation.

Despite the positive aspects cited and the rapid bioethanol expansion around the world, the concerns on its sustainability and challenges have increased. Among the drawbacks associated with bioethanol production are the high amount of fresh water consumed and the high generation of wastes, specifically the vinasse generated in the conventional processes (13 liter of vinasse/liter of ethanol produced).

One of the alternatives for overcoming these problems is the use of a continuous extractive process, which is associated to a higher productivity when compared to the traditional modes of operation adopted in the industrial sector. This system consists of three interconnected units: a fermentor, a membrane unit and a flash vessel under vacuum. The process is attractive for removing ethanol simultaneously to its production, reducing product inhibition in yeasts, resulting in high performance, productivity and non-necessity of heat exchangers.

Another point to be overcome in the industrial sector is the lack of robustness of the fermentation in the presence of fluctuations in operational conditions, which leads to changes in the kinetic parameters, with impact on yield, productivity and conversion. These changes are very common in plants of alcoholic fermentation.

Through the implementation of an extractive process associated with a robust model (under parameter re-estimation) to be used in optimization procedures and development of controllers, the performance of the fermentative processes is much improved, with gains of productivity, less waste generation and less water consumption.

In this chapter the advantages of the continuous extractive process and its kinetics are discussed through intrinsic modeling, considering substrate, ethanol and biomass inhibitions, rate of loss of viability of cells and temperature influence on the kinetics. This modeling reduces costs by allowing the investigation of suitable conditions by simulation and the development of optimizators and controllers. Besides, this chapter presents the methodology of continuous re-estimation of the kinetic parameters of existing models, which when associated to sensitivity analysis reduces the complexity of this task to obtain robust models, which are accurate even under changes in operational conditions.

NOMENCLATURE

D	Dilution rate (h^{-1})
F_0	Fresh molasses flow rate (m^3/h)
F	Fermentor outflow rate (m^3/h)
F_P	Permeate flow rate (m^3/h)
F_{PU}	Fermentor purge flow rate (m^3/h)
Fr	Flash tank liquid outlet flow rate (m^3/h)
F_{re}	Return stream flow rate (m^3/h)
F_V	Flash tank vapor outlet flow rate (m^3/h)
K_{dp}	Coefficient of death by ethanol (m^3/Kg)
K_{dT}	Coefficient of death by temperature (h^{-1})
K_i	Substrate inhibition parameter (m^3/kg)
K_s	Substrate saturation parameter (kg/m^3)
m	Parameters of cellular inhibitions
m_p	Luedeking-Piret non growth- associated constant ($kg/[kg \cdot h]$)
m_x	Maintenance parameter ($kg/[kg \cdot h]$)
μ_{max}	Maximum specific growth rate (h^{-1})
n	Parameter of product inhibitions
P	Ethanol concentration (m^3/Kg)
P_{max}	Product concentration when cell growth ceases (kg/m^3)
P_V	Flash tank pressure (kPa)
P_0	Inlet product concentration (Kg/m^3)
Pr	Product concentration in the flash tank liquid outlet flow (Kg/m^3)
Pv	Product concentration in the flash tank vapor outlet flow (Kg/m^3)
r_d	Rate of unviable cells (Kg/m^3h)
rx	Rate of viable cell growth (Kg/m^3h)
rs	Rate of substrate utilization (Kg/m^3h)
RSD	Residual standard deviations (%)
r_P	Rate of products production (Kg/m^3h)

S_0	Inlet substrate concentration (Kg/m^3)
S	Fermentor substrate concentration (Kg/m^3)
Sr	Substrate concentration in the flash tank liquid outlet flow (Kg/m^3)
Sv	Substrate concentration in the flash tank vapor outlet flow (Kg/m^3)
t	Time (h)
V	Total volume of the system (m^3)
X	Biomass concentration (Kg/m^3)
X_d	Concentration of dead biomass (Kg/m^3)
X_{max}	Biomass concentration when cell growth ceases (kg/m^3)
X_P	Biomass concentration in the permeate flow (Kg/m^3)
X_t	Total cellular concentration (Kg/m^3)
X_V	Biomass viable concentration (kg/m^3)
Y_{px}	Luedeking-Piret growth associated constant (kg/kg)
Y_x	Limit cellular yield (kg/kg)

1. Introduction

With the requirement of new energy sources resulting from the advent of industrial capitalism and the rapid growth of population, several alternative energy sources were exploited for centuries without the concerns on sustainability. In the last 15 years, with the increased concerns about climate changes, bioenergy gained importance in the world scenario, and nowadays biomass contributes for the supply of a significant amount of energy in developed countries: 4% in the USA, 18% in Sweden and 20% in Finland, and is projected to achieve 30 % of the total quantity produced in the world by 2050 (Rosillo-Calle et al., 2000), or 50 % of the total energy demand in developing countries (Demirbas, 2008). Bioethanol production has achieved an exponential growth since 2000, and is presumed to grow at the same level in the future (Granda et al., 2007).

Bioenergy presents several advantages over other conventional sources in the environmental and socioeconomic aspects only when produced in a sustainable and efficient way. Among the benefits of bioenergy production and use are job creation, especially in agricultural sectors; reduction in the level of greenhouse gas emissions due to the CO_2 sequestration by cane cropping (Goldemberg et al., 2008); less pollutants generation compared to fossil sources on account of less sulfur and nitrogen contents in biofuels; security energy, obtained by renewable matter; and wide availability of well distributed sources around the world, different from fossil materials that are limited to specific regions (Demirbas, 2008).

The Brazilian bioethanol manufacturing has showed to be more sustainable, environmentally friendly and economically feasible than the ethanol from corn, as it consumes only a small amount of fossil fuels. Only trucks for transportation, agricultural machinery and fertilizers depend on fossils as raw materials (Granda et al., 2007). In addition, all the amount of CO_2 exhausted from vehicles (running on bioethanol) and burning of bagasse in boilers is absorbed by photosynthesis of sugarcane. The CO_2 generation has been depleted with the advent of mechanization and the resulting gradual elimination of the sugarcane burning practice that precedes harvesting, as well as with the gain in efficiency in industrial boilers, which makes the CO_2 balance more favorable when compared with fossil fuels (Miranda et al., 1994).

Despite all overall advantages of bioethanol and the advancements achieved in that area, the fossil fuels are still economically more competitive in many countries. In addition, many problems in the environmental sector must be solved, as the great volume of residues, especially vinasse generated by the current fermentation processes, and water consumption for diluting molasses. In these aspects, many efforts must be concentrated to change these scenarios in order to make bioethanol production more competitive and environmentally friendly.

Taking this into consideration, this chapter discusses the development and advantages related to the use of extractive processes for bioethanol production coupled to the use of suitable and robust models for improving the process performance in the industrial sector.

1.1. Vinasse Generation and Water Consumption in Fermentation Existing Processes

Vinasse is considered one of the main polluting byproducts of processes of alcoholic fermentation. This effluent produces several damages to the environment when discarded without treatment due to the high BOD (biochemical oxygen demand), around 30 – 60 g/L, low pH due to organic acids, high solids content, and high volumes.

The quantity of vinasse produced by common processes is about 12-15 L per liter of ethanol generated (Navarro et al., 2000). In 2007, only Brazil was responsible for the production of 18 million cubic meters of ethanol (Goldemberg et al., 2008), and with the rapid expansion in the ethanol sector, the volume of vinasse tends to increase, requiring new technologies for its use and treatment. In addition, new biotechnological processes must be developed in order to reduce this amount.

Among the existing methods for vinasse re-use are the thermal concentration, application in lands, methane production (Navarro et al., 2000), cattle feed and fertilizer (Morin-Couallier et al., 2008), fungi production, material of construction, incineration, methane production and recycle of vinasse to partially dilute molasses in fermentation (Braunbeck and Cortez, 2005).

The thermal concentration of vinasse is very energy demanding (Kannabiran and Pragasam, 1993) and does not eliminate completely the problem of its generation; the application in lands as fertilizer is limited by high costs of transportation and storage involved, and can be prohibited by long distances faced to distribute the product.

The use of vinasse for methane production was practiced by a few distilleries in Brazil, and this gas was used as fuel in trucks, but this technology showed to be non competitive with the oil prices, and high costs with maintenance of specific converted trucks also made this technique unusual (Braunbeck and Cortez, 2005).

The technologies discussed above in this chapter for the use of vinasse do not reduce fresh water consumption in distilleries, which is considered another problem that must be solved to make bioethanol production more sustainable. An ideal strategy would solve the consumption of water and the amount of effluent generated, simultaneously.

Among the improvements in bioethanol production achieved with the strategy proposed is the reduction of operational costs by maintaining the industrial plant under optimal operational conditions, which is possible through the use of robust models and controllers;

and also the decrease in effluent generation. Extractive processes have shown to be suitable for reducing costs and vinasse generation and present higher productivity compared to the traditional operational strategies used in the industrial sector, such as batch, fed-batch and continuous modes.

2. Extractive Processes for Ethanol Production

Despite the researchs in ethanol production, the existing processes in Brazil commonly run in fed-batch or continuous modes, which produce great quantities of vinasse and are subject to ethanol inhibition, reducing productivity. In order to overcome this drawback, some works are focusing on the development of extractive fermentation processes, which consists of interlinked units composed by a bioreactor (ethanol production), and an extractive unit (for ethanol removal), as depicted in Figure 1. Among the advantages of product removal are the reduction of the inhibitory effect by ethanol; less vinasse generation due to the possibility of feeding molasses at higher concentrations, which reduces costs in waste treatment; and the possibility of eliminating one distillation column. It is important to stress that distillation is considered one of the most energy demanding operations in the industrial sector. The ethanol separation during the course of fermentation maintains the product concentration below inhibitory levels.

Thus, an efficient configuration of extractive fermentative process, and the design of compact plants (with high productivity and lower consumption in distillation) requires studies on available technologies for ethanol removal and optimization of operational variables. The most commonly techniques used in extractive processes are pervaportation, gas stripping, extraction with solvent and flash separation under vacuum, as presented on the following sections of this chapter.

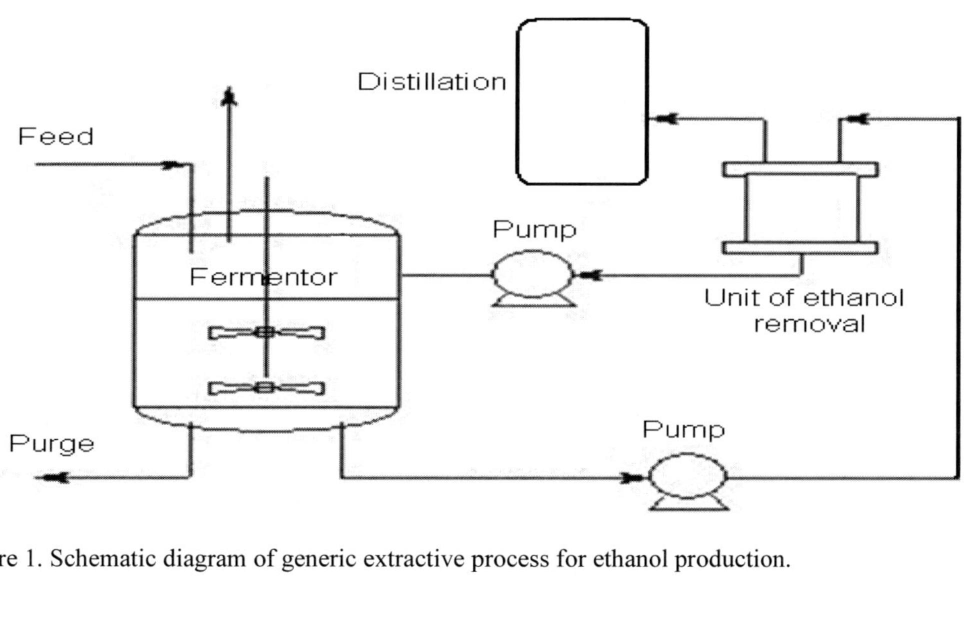

Figure 1. Schematic diagram of generic extractive process for ethanol production.

2.1. Gas Stripping

The gas stripping method for ethanol removal was evaluated by some authors, such as Taylor et al. (1996), and is based on the stripping of ethanol in a column with the injection of a gas in the bottom. The more volatile component is stripped from the mixture due to its high volatility. The recovery of bioethanol from the stream is performed by condensation. Among the carrier gases used in gas stripping processes are CO_2, O_2-free N_2 or H_2. Ezeji et al. (2004) evaluated gas stripping for ethanol-acetone-butanol production, achieving higher productivity than in batch modes of operation. For alcoholic fermentation, it is worthwhile mentioning that the anaerobic conditions must be maintained in the process, and the outlet gases of fermentation can be potentially used as carriers and be recycled, but studies of the effect of carrier gases on cellular metabolism are required. The ethanol obtained at the top of the column usually presents high concentration after condensation in a cooler at 350 Kg/m^3, which minimizes costs with downstream separations by distillation. The feeding of water in the reactor is required only to supply the losses due to the gas stripping, maintaining a constant total volume.

2.2. Solvent Extraction

The solvent extraction is one of the available techniques to reduce the inhibition by end product in ethanol fermentation, and is considered efficient energetically (Mariorella et al., 1984 and Daugulis et al., 1991) when compared to distillation. The project of the solvent extractive process is complex and the achievement of high productivity depends strongly on variables such as extraction efficiency, biocompatibility of the solvent (Cheng and Wang, 2007), volume required for separation, chemical stability, costs of separation of the product and solvent recovery, emulsion or foam formation in the broth, safety of the industrial plant, and environmental risk when this solvent is discharged. The solvent selectivity and losses must also be considered.

Some authors like Wang and Achenie (2002) have applied tools as computer-aided molecular design to search for a biocompatible solvent for extractive fermentation processes, and the substances evaluated were 3-hexanone, ethyl isopropyl ketone, sec-butyl isopropyl ether, and *n*-butyl acetate, which was found to be the most biocompatible solvent.

The modern solvent design techniques are based on computational algorithms, such as heuristic numeration, knowledge-based approaches, branch and bound approaches, mixed-integer optimization, genetic algorithm, simulating annealing and other methods (Wang and Achenie, 2002), and the problems are also subject to physical-chemical constraints.

In experimental works, the vegetable oils have been investigated as extractant agents to remove ethanol from fermentation broths. The oleyl alcohol has many desirable characteristics as a solvent, such as low toxicity to yeasts, but for application in the commercial sector its extractive power is not attractive. According to Mehta and Fraser (1985), molecules of high molecular weight, such as hydrocarbons and vegetable oils would be a potential option for ethanol extraction due to its capacity of forming conjugate pairs of molecules in the presence of ethanol. Some examples cited by the authors are hexadecane, cottonseed oil, olive, coconut and soybean oils. The main problem faced was the low extraction capacity obtained in solutions diluted in ethanol (Rahman et al., 1995).

In a recent work, it was found that some branched alcohols with a range of 14- 20 carbons have presented higher extractive performance when compared to other solvents as oleyl alcohol and 1- dodecanol (Offeman et al., 2008). This allows the extraction of ethanol from the fermentation broth even in lower concentrations. It was also shown that these alcohols with long chains are non toxic to microorganisms and present low solubility in the aqueous phase. On the other hand, alcohols with shorter chains (range of 2 to 10 carbons) are toxic to microorganisms and consequently not suitable for extractive processes for ethanol production.

The same authors (Offeman et al., 2008) also evaluated experimentally the effect of the toxicity of alcoholic solvents on the yeasts. The C8-C12 alcohols presented inhibitory or toxic effect on ethanol production, while C14- C20 alcohols were superior for presenting low solubility in the aqueous phase, which results in a low influence in the yeast metabolism and productivity.

2.3. Pervaporation

Extractive processes by pervaporation apply a membrane for selective removal of ethanol from the fermentation broth. By one side, the broth with diluted ethanol is fed, and on the other side, a vacuum pump is installed to produce low pressures, which forces ethanol to pass through the membrane. A trap at low temperature for condensing and recovering the ethanol is installed in the system. The main advantage of pervaporation is that this technique enables the concentration of ethanol in the permeate flow above the azeotropic composition, requiring one stage of separation, which do not occur in other extractive processes that are limited by the thermodynamic equilibrium. Also, the pervaporation processes run at environmental temperature, which do not require additional energy expenses, as in distillations.

Through the use of a silicate membrane, Nomura et al. (2002) successfully obtained a separation factor much higher than polymeric membranes, as investigated by Nakao et al. (1987), and also an ethanol high permselectivity. Additionally, the salt effect of the fermentation broth increases the ethanol selectivity when compared to the mixture water-ethanol.

2.4. Extractive Processes under Vacuum

The extractive process under vacuum is remarkable to be simpler to operate and build up comparing to other systems, as it requires only a tank linked to a vacuum pump and condenser. Thus, the problems with membrane fouling, solvent separation, toxicities on the cells and emulsification are eliminated. This model is also easier to adapt to existing industrial processes in Brazil. In addition, the evaporation of ethanol removes heat from the system, which potentiality eliminates the necessity of heat exchangers for cooling the fermentor and saves utilities consumption when the flash is operated at 33ºC. According to Andrietta (1994), the heat exchangers are one of the most expensive equipments in the ethanol sector.

This process was built up and experimentally validated for bioethanol production in bench scale by Atala (2004), as cited by Mariano et al. (2008). The medium used was sugarcane molasses at high concentrations, from 180 to 330 Kg/m^3 of TRS (total reducing sugar), and the microorganism *Saccharomyces cerevisiae* at 30 Kg/m^3 after the achievement of steady state. The flash vessel is able to maintain ethanol concentration at the broth medium bellow the inhibitory level (around 40 - 60 Kg/m^3). These characteristics of operation guarantee less vinasse formation for allowing the feeding of molasses at higher concentration than in common existing processes, and increases ethanol productivity to 10 $Kg/m^3.h$.

3. Reduction of Operational Costs by Robust Modeling

One of the challenges in bioethanol production is the reduction of operational costs. One alternative to achieve this goal is the use of extractive processes and the maintenance of the operational unit under optimal conditions to reach high productivity, yield and conversion, with less effluent generation and fresh water consumption. This can be done using simulation tools and efficient optimizators and controllers, which requires robust and well developed models.

Although many works have been published about bioethanol modeling (Phisalaphong, et al. 2005; Andrietta, 1994; Atala, 2004 and Andrade et al. 2007), the kinetic parameters of the models change with operational conditions such as yeast lineage, characteristics of the raw material used, presence of contaminants in fermenters, microbial load in sugarcane and other characteristics of the process used to construct the model.

It is important to stress that the presence of contaminants in raw material in Brazil is not eliminated by sterilization due to high costs of operation that would economically prohibit bioethanol production.

The changes in the operational conditions mentioned above alter drastically the kinetics of fermentation. Estimation of model parameters is one of the important but difficult problems in the construction of a robust model. Since the number of model parameters tends to increase, their estimation become a difficult and time-absorbing step. Furthermore, as model parameters vary according to the fermentation conditions in bioprocesses, it is necessary to re-estimate these parameters. In order to save the effort and time, sensitivity analysis can be applied for parameter screening. The most significant parameters are re-estimated when changes in fermentation conditions occur, obtaining an accurate model.

3.1. Source of Variations in Fermentative Processes

The variations in operational conditions occur frequently in bioethanol production, and affect the kinetic of fermentation. Some sources of variations can be cited, such as different yeasts lineages of *S. cerevisiae*, the variability of sugarcane due to the type of soils, climatic conditions, pluviometric precipitation, maturity of the raw material, and post harvest conditions of storage, microbial load and others.

Andrade et al. (2009) discuss that some authors like Phowchinda and Strehaiano (1999) have investigated the impact of different sugars on the overall kinetic of fermentation by *S. cerevisiae*. The rate of sucrose uptake is lower than glucose and fructose. The kinetic rate is also influenced by nitrogen source in yeasts (Andrade et al., 2009) and heavy metals in the molasses, although these effects have not been added as terms of kinetic expression in the modeling of alcoholic fermentation.

For batch mode in alcoholic fermentation by *S. cerevisiae*, Andrade et al. (2007) have shown that the re-estimation of 4 parameters of a total of 11 is sufficient for obtaining a robust model. The model used by the authors is formed by three differential expressions (Eq. 1-3) as follows:

$$\frac{dX}{dt} = \mu_{max} \frac{S}{K_s + S} \exp(-K_i S)(1 - \frac{X}{X_{max}})^m (1 - \frac{P}{P_{max}})^n X \quad (1)$$

$$\frac{dS}{dt} = Y_{px} r_x + m_p X \quad (2)$$

$$\frac{dP}{dt} = (r_x / Y_x) + m_x X \quad (3)$$

where μ_{max} is the maximum specific growth rate (h^{-1}); K_s the substrate saturation parameter (kg/m^3); K_i is the substrate inhibition parameter (m^3/kg); X_{max} is the biomass concentration when cell growth ceases (kg/m^3); P_{max} is the product concentration when cell growth ceases (kg/m^3); *m* and *n* are parameters of cellular and product inhibitions; Y_{px} is Luedeking-Piret growth associated constant (kg/kg); m_p is the Luedeking-Piret non growth- associated constant (kg/[kg·h]); Y_x and m_x are the limit cellular yield (kg/kg) and maintenance parameter (kg/[kg·h]), respectively. In these equations, *X* is the biomass, *S* the substrate, and *P* the ethanol concentration (m^3/Kg).

The parameters μ_{max}, X_{max}, P_{max}, Y_x and Y_{px} are temperature dependent. The authors show the re-estimation of these parameters is sufficient to obtain a model with a good capacity of prediction under changes in raw material quality, maintaining the remaining ones, K_s, K_i, *m*, *n*, m_p, m_x, fixed. The reduction of the number of parameters to be re-estimated makes model updating quick and simple, especially for models with great number of parameters, as structured models, or complex models that consider additional equations on their kinetics.

The main advantage of parameter updating is the obtaining of an accurate model to be applicable in simulations, optimizations and control, which increases the performance of the fermentative process, reducing costs of production for maintaining it under optimal operational conditions. A non-valid model leads to errors in decision making.

In a recent work, Andrade et al. (2009) discuss and propose an algorithm for screening and re-estimating the kinetic model parameters for alcoholic fermentation based on the well known Plackett-Burman sensitivity analysis, and quasi-Newton method. In a first step, kinetic parameters of a model are selected by sensitivity analysis. The authors also suggest that Plackett- Burman is superior to one-at-a-time sensitivity analysis because all the parameters are simultaneously evaluated in the first technique. In a second step, only the most relevant parameters are updated when changes in operational conditions occur.

3.2. Plackett-Burman (PB) Methodology for Screening of Kinetic Parameters for Re-Estimation Problem

Plackett-Burman design can be applied to evaluate the influence of a great number of factors on some given responses (Barros Neto *et al.*, 1996). With a PB design of 20 trials, a maximum of 19 variables can be analyzed. A lower number of factors can be used to allow for degrees of freedom for error calculation. The main idea of this technique is to screen the kinetic parameters that have significance on the behavior of the process, and the non-significant ones are maintained constant during the parameters re-estimation problem. Re-estimation is performed in order to represent the kinetics of the process at different conditions of molasses composition and microorganism strain.

In the work of Andrade et al. (2009), Plackett-Burman design was used to investigate the influence of kinetic parameters (11) on the responses: fermentation time (t), product concentration at the end of fermentation (P) and cell concentration at the end of fermentation (X). For this purpose, parameters of a deterministic model at 34° C were tested at two levels, high (+) and low (-), according to Table 1. A number of 20 trials was chosen and the calculations were made with 8 degrees of freedom.

Through simulation using FORTRAN 90, the values of t, X and P were obtained for each fermentation and used as responses in PB technique. Initial conditions of concentrations of biomass, substrate and product were fixed at 0.7, 119.1 and 3.9 Kg/m^3, respectively.

The software STATISTICA, version 7.0, was used for the screening of the parameters that have higher influence on the kinetic behavior, as shown in Table 2. Values of effects on responses were estimated at the 95 % confidence level. Results presented in bold are significant.

According to Table 2, μ_{max} and n were statistically significant for fermentation time, m_x and m_p for product concentration at the end of fermentation. The effects of P_{max}, Y_x and Y_{px} were observed on the model for all responses (t, P and X).

Table 1. high (+) and low (-) levels of parameters for PB technique

Parameter	Level -1	Level +1
μ_{max}	0.35	0.42
X_{max}	38.7	47.3
P_{max}	67.5	82.5
Y_x	0.0356	0.0436
Y_{px}	8.9393	10.9258
K_s	3.69	4.51
K_i	0.0036	0.0044
m_x	0.18	0.22
m	0.9	1.1
n	1.35	1.65
m_p	0.09	0.11

Andrade et al. (2009)

Table 2. Effect estimates on time of fermentation, final ethanol and biomass concentration, from PB design. Values in bold are considered significant.

	Effect on t	Effect on X	Effect on P
μ_{max}	**-2.1610**	0.0464	0.1400
X_{max}	-0.1270	-0.0094	0.0241
P_{max}	**-3.7370**	**0.1193**	**-0.4875**
Y_x	**3.8330**	**0.7058**	**9.0418**
Y_{px}	**3.6630**	**-0.1133**	**8.9003**
K_s	-0.1730	0.0096	-0.1873
K_i	1.0650	-0.0306	-0.1860
m_x	0.6730	-0.0422	**-0.8816**
m	-0.0870	0.0185	-0.1092
n	**1.7430**	-0.037918	-0.1195
m_p	-0.0570	-0.0093	**1.0309**

Andrade et al. (2009).

The parameters that are considered non-significant are fixed in the re-estimation problem. In order to evaluate the importance of re-estimation of parameters, the authors simulate a mechanistic model with parameters obtained by Andrade et al. (2007) and perform some new batch fermentations at different molasses (different harvesting) and yeast strain conditions. As a result, the original model presents poor capacity of prediction.

Thus, the authors propose the re-estimation of the most significant parameters of Table 2 and the less significant parameters were maintained fixed as X_{max}=43, K_s=4.1, K_i=0.004, m_p=0.1, m=1.0, n=1.5 and m_x=0.2. In this methodology, only 4 parameters were re-estimated: μ_{max}, P_{max}, Y_x and Y_{px}. A mechanistic and robust model is obtained, with good quality of prediction for different experimental condition.

The residual standard deviation (RSD), Eq. (4), was used for characterizing the quality of the model prediction. After the re-estimation procedure, it was shown the original model achieved good accuracy (deviations decreased from about 35 to 10 %).

$$\mathrm{RSD}(\%) = \left(\frac{\sqrt{\mathrm{RSD}}}{\bar{d}_p}\right)100 \tag{4}$$

where $\mathrm{RSD} = \frac{1}{np}\sum_{p=1}^{np}(d_p - x_p)^2$, in which x_p and d_p are, respectively, the value predicted by the mathematical model and experimental value, $\bar{d}_p$ is the average of the experimental values and *np* is the number of experimental points.

3.3. Mathematical Modeling of an Extractive Process under Vacuum

The mathematical modeling of the extractive process potentially reduces costs of production by allowing the operation under suitable conditions. This modelling will consider the non-structured approach, where cells are represented as a single component in the mixture, according to Bailey et al. (1986). The non- segregated approach was also considered, or in other words, the average cell properties are used.

According to Phisalaphong et al. (2005), the mathematical modeling of the process establishes a relation between kinetic and environmental parameters, usually represented by a modified Monod type model. This model must represent the substrate and product inhibitions, considering that the lineages used in industrial processes are subject to these effects on cellular growth.

Figure 2 presents a continuous extractive process for ethanol production coupled to a flash vessel with cell recycle. The modeling of the process takes into account the volume occupied by the cells, intracellular ethanol and terms of cell death, considering that continuous processes are operated for long periods. The term ρ, defined by Monbouquette (1992), is a relation between dry cell mass and the volume of wet cells, and γ, the relation between the inter and extracellular ethanol concentrations.

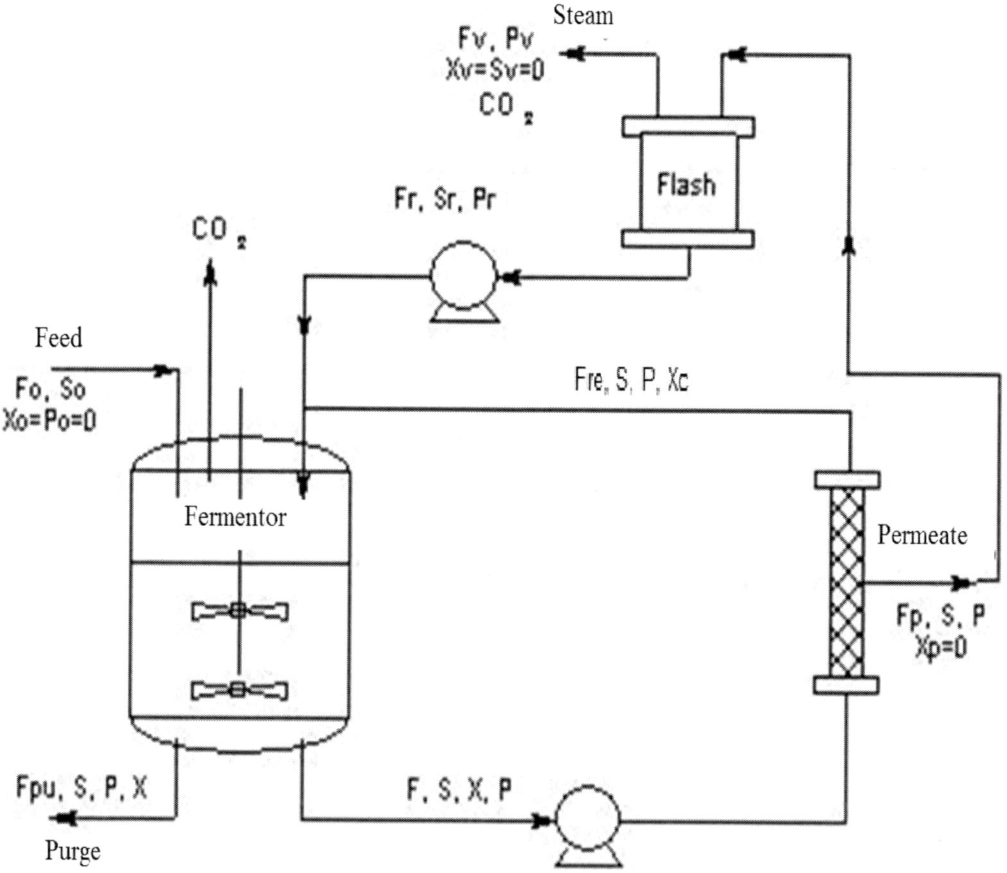

Figure 2. General scheme of extractive process using a flash vessel.

For the modeling of the extractive process under vacuum, some considerations are made, such as a fermentative system represented by a well stirred tank reactor; the total volume of the fermentative process remains constant; the biomass content is removed from the system only in the purge stream, or in other words, the membrane retains completely the cells; the feed stream contains only diluted sugarcane molasses, with absence of cells and ethanol.

Thus, the equations of mass balance in the fermentative system are represented by:

- Viable biomass

Accumulation = in – out + generation term

$$\frac{d(Xv \cdot V)}{dt} = rx \cdot V - rd \cdot V - F_{PU} \cdot Xv \tag{5}$$

Equation (5) can be written in a simpler way, by:

$$\frac{dX_V}{dt} = rx - rd + \phi \cdot D \cdot Xv \tag{6}$$

$$\phi = \frac{F_{PU}}{F_O} \tag{7}$$

$$D = \frac{F_O}{V} \tag{8}$$

The symbol ϕ denotes the purge rate of the system, and D the dilution rate, defined by Eqs (7) and (8), respectively.

- Cellular death

Considering that the continuous fermentation occurs for long periods, the term of cell death must be taken into account in the modeling:

Accumulation = in – out + generation

$$\frac{d(Xd \cdot V)}{dt} = rd \cdot V - F_{PU} \cdot Xd \tag{9}$$

Thus:

$$\frac{dX_d}{dt} = rd + \phi \cdot D \cdot Xd \tag{10}$$

- Product formation

Accumulation = in – out + generation

$$\frac{d(P \cdot V)}{dt} = -[F_{PU} \cdot P + F_V \cdot P_V] + rp \cdot V \tag{11}$$

Considering that a fraction of ethanol is intracellular:

$$\frac{d(P \cdot V)}{dt} = \frac{d}{dt}[(vol.\ reac. - vol.\ cel.) \cdot P + [P\ intra\ cel.] \cdot V_{Cel}\ ] = \frac{d}{dt}\left[\left(1 - \frac{Xt}{\rho}\right) \cdot P \cdot V + \frac{Xt}{\rho} \cdot \gamma \cdot P \cdot V\right] \tag{12}$$

$$\frac{d}{dt}\left[\left(1 - \frac{Xt}{\rho}\right) \cdot P \cdot V + \frac{Xt}{\rho} \cdot \gamma \cdot P \cdot V\right] = rp \cdot V - F_O \cdot [(\phi) \cdot P + (1-\phi) \cdot P_V] \tag{13}$$

$$\frac{dP}{dt} = \frac{rp - D \cdot [(\phi) \cdot P + (1-\phi) \cdot P_V] + \frac{(1-\gamma)}{\rho} \cdot P \cdot (rx - \phi \cdot D \cdot X_{Total})}{\left[1 + \frac{X_{Total}}{\rho} \cdot (\gamma - 1)\right]} \tag{14}$$

the symbol ϖ is the rate of permeate of the system, and is defined by Eq. (15):

$$\varpi = \frac{F_P}{F_O} \tag{15}$$

- Substrate consumption

Accumulation = in – out + generation

$$\frac{d(S \cdot V)}{dt} = F_O \cdot S_O - [F_{PU} \cdot S] - rs \cdot V \tag{16}$$

Considering that the concentration of substrate in the interior of cell is zero:

$$\frac{d}{dt}\left[\left(1 - \frac{Xt}{\rho}\right) \cdot S \cdot V\right] = F_O \cdot [S_O - S \cdot (\phi)] - rs \cdot V \tag{17}$$

Writing Eq. 17 in a simplest way: $\frac{d}{dt}\left[\left(1 - \frac{Xt}{\rho}\right) \cdot S\right] = D \cdot [S_O - S \cdot (\phi)] - rs$ (18)

and finally

$$\frac{dS}{dt} = \frac{D \cdot [S_O - S \cdot (\phi)] - rs + \frac{S}{\rho} \cdot [rx - \phi \cdot D \cdot X_{Total}]}{1 - \frac{X_{Total}}{\rho}} \tag{19}$$

The symbols Xv, Xd, S and P are the concentrations of viable and dead cells, substrate (total reducing sugar) and product, respectively; So denotes the substrate concentration in the feed stream and D is the dilution term (D=Fo/V, where Fo is the volumetric flow rate in the feed stream and V is the total volume of the fermentative system).

In the balance of product and substrate the volume occupied by biomass was considered. According to Monbouquette (1992), for cultivation at high cellular densities where the fractions of cell are greater than 10 % of the reactor volume, the intrinsic approach must be considered.

The specific rates of biomass formation, cell death, product formation and substrate uptake are given by the Eqs. (20) - (23), respectively:

$$rx = \mu_{\max} \cdot \left(\frac{S}{Ks + S} \right) \cdot \exp\left(- Ki \cdot S \right) \cdot \left(1 - \frac{P}{P_{\max}} \right)^{n} \cdot \left(1 - \frac{X_t}{X_{\max}} \right)^{m} \cdot X_V \tag{20}$$

Eq. (20) is based on the expression proposed by Levenspiel (1980) and was used by Andrietta (1994) and recently by Atala et al. (2001) and Andrade et al. (2007) for modeling of ethanol fermentation processes, and presented a good agreement with experimental data.

The use of Eq. (20) is suitable for modeling this process that run at high cellular concentration as it presents a term of growth inhibition caused by existing yeasts on the fermentor. This occurs due to the difficulty of oxygen and substrate transfer in the fermentative broth at high yeasts concentration, which is common in processes with cell recycle.

The same expression also considers the substrate as a limiting factor of growth and as inhibitor, which occurs with sugarcane molasses at high concentration due to the presence of components that reduce the microbial growth.

$$rd = \left(K_{dT} \cdot \exp(Kdp \cdot P) \right) \cdot Xv \tag{21}$$

$$rp = Ypx \cdot rx + mp \cdot Xv \tag{22}$$

Eq. 22 is suitable for alcoholic fermentations, considering that the expression takes into account a term for ethanol formation associated with yeasts growth (r_x) or existing viable microorganisms (X_v).

$$rs = \frac{rx}{Yx} + mx \cdot Xv \tag{23}$$

Eq. (23) takes into account the substrate consumption for cellular growth, microbial activities, mobility, and transport of solutes by the cellular membrane.

Overall Mass Balance in Extractive Process

Some considerations must be taken into account in the overall mass balance of the extractive process, as presented:

- the biomass concentration in the permeate is zero (stream F_p), occurring total separation of yeasts by the membrane system;
- the concentrations of ethanol and sugars in inlet and outlet streams of filtration system are held constant;
- the biomass concentration in the fermentor does not change during the time of operation;

The cellular concentration in the reactor is held constant due to the use of a purge (stream F_{PU}), which allows the withdrawal of metabolites of fermentation that concentrate in the system and are not removed by flash distillation under the working conditions. The purge also renews the cells continuously, according to Andrietta (1994).

When a mass balance is applied to the process of Figure 2, the following expressions are obtained:

$$F_F = F_O + F_R + F_{RE} = F_{PU} + F \tag{24}$$

$$F_R = F_P - F_V \tag{25}$$

$$F_P = F - F_{RE} \tag{26}$$

$$S_F = \frac{F_R S_R + F_O S_O + F_{RE} S_{RE}}{F_F} \tag{27}$$

$$F_0 = F_V + F_{PU} \tag{28}$$

$$F_V = F_O \cdot (1 - \phi) \tag{29}$$

$$t_r = \frac{V}{F} \tag{30}$$

$$V = V_{reactor} + V_{filter} + V_{pipes\_of\_cell\_recycle} \tag{31}$$

The residence time is represented by (tr) and (V) is the volume occupied by all the regions were reactions occur.

Conclusion

This chapter brings a review of existing available techniques for extracting ethanol from diluted broth. By its simplicity of installation and operation, the extractive process under vacuum was presented as a remarkable and suitable methodology to be adapted to industrial plants, especially in Brazil. Although other techniques, such as pervaporation, are more efficient by the higher concentration of ethanol obtained with one stage, the extraction under vacuum coupled to existing distillation towers probably requires less investments.

The use of the extractive process under vacuum for bioethanol production, with the suitable model presented in this chapter, with robust optimizators and controllers that act continuously when changes in fermentation conditions occurs, potentially enhances ethanol production.

References

Andrade, R.R.; Rivera, E.C.; Atala, D.I.P.; , Maciel Filho, R.; Maugeri Filho, F.; Costa, A.C. (2009). Study of kinetic parameters in a mechanistic model for bioethanol production through a screening technique and optimization. *Bioprocess. Biosyst. Eng.* [doi 10.1007/s00449-008-0291-8].

Andrade, R.R.; Rivera, E.C.; Costa, A.C.; Atala, D.I.P.; Maugeri Filho, F; Maciel Filho, R. *Appl. Biochem. Biotechnol.* 2007, *136-140*, 753.

Andrietta, S. R. *Modelagem, simulação e controle de fermentação alcoólica em escala industrial;* (PhD. Thesis) Unicamp: Campinas - SP, 1994; 178p.

Atala, D. I. P. *Montagem, Instrumentação, Controle e Desenvolvimento Experimental de um Processo de Fermentação Alcoólica Extrativo;*(PhD. Thesis) Unicamp: Campinas – SP, 2004.

Bailey, J.E; Ollis, D.F. *Biochemical Engineering Fundamentals;* Second Edition; Mc Graw-Hill: New York, NY, 1986; Vol. 1, pp 984.

Braunbeck, O.A. Cortez, L.A.B. In *The use of biomass for energy production in Brazilian industries. (in Portuguese)*; Rosillo-Calle, F.; Bajay, S.V.; Rothman, H.; Editora Unicamp: Campinas, SP, 2000; Vol. 1, pp 215-246.

Cheng, H.; Wang, F. *Chem. Eng. Sci.* 2007, *62*, 4316 – 4324.

Daugulis, A.J.; Axford, D.B.; McLellan, P.J. *Can. J.Chem. Eng.* 1991, *69*, 488–497.

Demirbas, A. *Energ. Conv.Manag.* 2008, *49*,2106-16.

Ezeji, T.C.; Qureshi, N.; Blaschek, H.P. *Appl. Microbiol. Biotechnol.* 2004, *63*,653–658.

Goldemberg, J.; Coelho, S.T.; Guardabassi, P. (2008). The sustainability of ethanol production for sugarcane. *Energy Policy* [doi:10.1016/j.enpol.2008.02.028].

Granda, C.B.; Zhu, L.; Holtzapple, M.T. (2007). Sustainable liquid biofuels and their environmental impact. *Environmental Progress* [doi: 10.1002/ep], Vol. 26.

Kannabiran, B.; Pragasam, A. *Geobios.* 1993, *20*, 108-12.

Levenspiel, O. *Biotechnol. Bioeng.* 1980, *22*, 1671-1687.

Maiorella, B.L.; Blanch, H.W.; Wilke, C.R. *Biotechnol. Bioeng.* 1984, *26*, 1003–1025.

Mariano, A. P.; Angelis, D. F.; Maugeri Filho, F.; Atala, D.I.P.; Maciel, M.R.W.; Maciel Filho, R. *Chem. Product. Process. Model.* 2008, *3*, 1-14.

Mehta, G.D.; Fraser, M.D. *Ibi.* 1985, *24*, 556–560.

Miranda, E.E.; Dourado, A.J.; Assunção, J.V. *Doenças respiratórias crônicas em quatro municípios paulistas (Chronic respiratory disease in four São Paulo municipalities).* Ecoforça: São Paulo - SP, 1994.

Monbouquette, H. G. *Biotechnol. Bioeng.* 1992, *39*, 498-503.

Morin-Couallier, E.; Fargues, C.; Lewandowski, R.; Decloux, M.; Lameloise, M. *J. Clean.Prod.* 2008; *16*, 655-663.

Navarro, A.R.; Sepúlveda, M. del C.; Rubio, M.C. *Waste manag.* 2000, *20*, 581-585.

Nakao, S.; Saitoh, F. ; Asakura, T; Toda, K; Kimura, S. J. *Membr. Sci.* 1987, *34*, 273.

Nomura, M.; Bin, T; Nakao, S. *Sep.Purif. Technol.* 2002, *27*, 59–66.

Offeman R.D.; Stephenson, S.K.; Franqui, D.; Cline, J.L. *Sep.Purif. Technol.* 2008, *63*, 444-451.

Phisalaphong, M.; Srirattana, N.; Tanthapanichakoon, W. *Biochem. Eng. J.* 2005, *28*, 36-43.

Phowchinda, O; Strehaiano, P. *Thammasat Int J Sci Technol.* 1999, *4*,2.

Rahman, M.A.; Rahman, M.S.; Asadullah, M. *Indian J. Chem.Technol.* 1995, *2*, 90–92.

Rosillo-Calle, F.; Bajay, S.V.; Rothman, H. (2000). *The use of biomass for energy production in Brazilian industries. (in Portuguese)*; Editora Unicamp: Campinas, SP, 2000; Vol. 1, pp 25-67.

Taylor, F.; Kiurantz, M. J.; Goldberg, N.; Craig, J.C. *Biotechnol. Bioeng.* 1996, *51*, 33-39.

Wang, Y.; Achenie, L.E.K. *Fluid Phase Equilibria.* 2002, *201*, 1–18.

In: Bioethanol: Production, Benefits and Economics
Editor: Jason B. Erbaum
ISBN 978-1-60741-697-5

Chapter 4

EXTREMELY THERMOPHILIC ENZYMES: THEIR UTILIZATION IN AGRICULTURAL WASTE CONVERSION FOR BIOETHANOL PRODUCTION

***Alessandra Morana*[1], *Luisa Maurelli*[1], *Francesco La Cara*[1] and *Mosè Rossi*[2]**

[1]Istituto di Biochimica delle Proteine, CNR, via P. Castellino 111, 80131 Naples, Italy

[2]University of Naples "Federico II", Dipartimento di Biologia Strutturale e Funzionale, Complesso Universitario Monte S. Angelo, via Cinthia, 80126 Naples, Italy

ABSTRACT

The conversion of lignocellulosic biomass to fermentable sugars for low-cost fuel production represents a major challenge in global efforts to utilize renewable resources rather than fossil fuels. However, lignocellulosic materials, which represent the most abundant polysaccharides in nature, have not encountered significant utilization because of the difficulties associated with their degradation. In such kinds of biomass, the chains of carbohydrates, mainly cellulose and hemicelluloses, are embedded in a lignin matrix, which hinders their efficient degradation. Thermal, chemical, and biochemical approaches have been proposed, both individually and in combination, as promising solutions to overcome this difficulty, but the drastic conditions required by many pretreatment techniques give rise to problems when using conventional enzymes in the following saccharification step. In this context, enzymes isolated from microorganisms growing at extreme temperatures could represent an attractive solution since they are thermostable, active at high temperatures (thermophilic) and resistant to solvents and detergents. These unusual properties make them interesting candidates for the development of saccharification processes of lignocellulosic biomass. Here, we report on the hydrolysis of agricultural waste by thermostable enzymes isolated from extremophilic microorganisms. In particular, polysaccharide-degrading activities from the hyperthermophilic archaeon *Sulfolobus solfataricus*, which grows optimally at acidic pH (2.0–4.0) and high temperatures (80-87°C) and originally isolated from a solfataric field in the area of Naples, have been used to hydrolyze at high temperature agro-based raw materials such as brewer's spent grains and corn stover. The raw material has been subjected to enzyme hydrolysis directly or after preliminary pretreatments in order to compare the efficiency of the enzyme saccharification process on differently treated material. Initial results indicated that incubation of the brewer's spent grains with a combination of glycolytic enzymes from *S. solfataricus* at high temperature (80°C) and acidic pH (5.0) provided glucose, arabinose and xylose as final products.

Keywords: Brewer's spent grain, Cellulose, Corn stover, Hemicellulose, Lignocellulosic biomass, Saccharification, Pretreatment, *Sulfolobus solfataricus*.

List of Abbreviations

CS	corn stover
BSG	brewer's spent grains
RBB-xylan	Remazol Brilliant Blue R-D-xylan
*p*NPX	*p*-nitrophenyl-β-D-xylopyranoside
*p*NPG	*p*-nitrophenyl-β-D-glucopyranoside
*p*NPA	*p*-nitrophenyl-α-L-arabinofuranoside
AZO-CMC	AZO-carboxymethylcellulose
HMF	hydroxymethylfurfural
LDPs	lignin degradation products

Introduction

During the last decades, considerable attention has been addressed to the conversion of starch-based materials, such as sugarcane and corn, into fuel ethanol and additional useful by-products [1, 2]. However, the "second generation" production of bioethanol derived from the utilization of cheaper substrates such as agronomic residues, also termed lignocellulosic biomass, could make bioethanol more competitive with respect to fossil fuels, and also solve the disposal problems of these matters [3, 4]. Lignocellulosic biomasses are an attractive feedstock for producing liquid fuels and chemicals because they are the most abundant source of carbohydrates on earth and are renewable and not in competition with food sources [5-7].

However, the commercial development of bioethanol from lignocellulose is hindered by several technical restrictions. Processing costs are high and mainly due to the incidence of the biomass and enzyme costs, and the technological solutions must be focused on maximizing the exploitation of all the biomass components, and on the achievement of high yields in each process step. Since the major components of lignocellulosic biomass are cellulose and hemicellulose besides lignin, the utilization of both cellulose and hemicellulose for bioethanol production is an ambitious goal that would make the process more convenient [8]. One promising technology for conversion of lignocellulosic biomass into ethanol is an enzyme-based process that utilizes a preliminary step to enhance the enzymatic conversion of the polysaccharides into fermentable sugars [9-11]. The results of studies on bioconversion of lignocellulosic materials (i.e., agricultural waste, woods, urban lignocellulosic waste) clearly indicates that the removal of lignin and the following increase of porosity can significantly improve the hydrolysis of the lignocellulose since the structural features of the material and the presence of lignin hinder the biomass bioconversion [12-15]. More in detail, the bioconversion process necessary to produce fuel ethanol from lignocellulose requires the three following steps: 1) *pretreatment*: delignification to release cellulose and hemicellulose from their complex with lignin, making more accessible these polysaccharides to the enzymes; 2) *saccharification*: conversion of the polysaccharide chains into fermentable sugars; 3) *fermentation*: fermentation, by yeasts or other appropriate microorganisms, of the resulting mixture of hexose and pentose sugars to produce ethanol.

Pretreatment methods may involve physical, chemical, biochemical and microbial approaches, alone or in combination, but the drastic conditions required by many of them (high temperature, pressure, low pH) give rise to problems when using mesophilic enzymes in the subsequent saccharification step. In order to overcome this difficulty, microorganisms thriving in habitats characterized by extreme conditions (high temperatures, low or high pH) can be taken under investigation as sources of polysaccharide-degrading enzymes, since they allow biotransformation processes at "non-conventional conditions" under which common enzymes are completely denatured. Thermophilic enzymes are expected to be a powerful tool in industrial biotransformation processes that run at harsh conditions, especially for the conversion of complex polymers. In fact, the possibility of utilizing elevated operational temperatures allows a higher rate of substrate conversion, better solubility of the substrates (particularly remarkable if the substrates are viscous or polymers), and absence of microbial contamination in the reaction media [16].

The unique stability of these biocatalysts at elevated temperatures (up to 110°C), extremes of pH and high pressure (up to 1000 bar) makes them a useful tool in the development of bioethanol production processes from lignocellulosic biomass [17, 18]. As an example, it is expected that material pretreated by steam explosion, will be more susceptible to the enzyme action at high temperatures, simplifying the general bioconversion process and also reducing the production costs. The polysaccharides cellulose and xylan (the main component of hemicellulose) contained in lignocellulosic residues are the main products derived from agricultural waste and represent a cheap source of feedstock that could have an enormous economic potential for the conversion of plant biomasses into chemicals and fuels. For full degradation of cellulose and xylan into fermentable sugars, the action of several enzymes (cellulase, β-glucosidase, xylanase, α-arabinosidase, β-xylosidase, acetylxylan esterase and α-glucuronidase) acting in cooperation is required. Cellulolytic and hemicellulolytic degrading enzymes from thermophilic microorganisms comprising archaea and bacteria have been investigated from *Acidothermus cellulolyticus*, *Clostridium thermocellum* and *Caldocellum saccharolyticum* [19-22]. In addition, hyperthermal cellulases and hemicellulases have been isolated and studied, and most of these enzymes have optimal temperature around or over 90°C [23-26]. Thermostable cellulases and xylanases have also been detected in different Bacillus, Alicyclobacillus and Sulfolobales species [27-31].

Here we report on the accomplishment of agricultural waste conversion processes by using thermostable and pH resistant cellulolytic and hemicellulolytic enzymes from the archaeon *Sulfolobus solfataricus*, a hyperthermophilic microorganism, isolated from a solfataric field in the area of Naples (Italy), and growing at acidic pH (2.0–4.0) and high temperatures (80–87°C) [32]. A mix of enzymes from *S. solfataricus* strain Oα, previously isolated in our lab for its ability to grow on minimal medium supplemented with xylan as a unique carbon source [31], were used to degrade at high temperature the polysaccharidic fractions contained in corn stover (CS) and brewer's spent grain (BSG) selected as the biomass [33-35]. The enzyme cocktail contained all the main glycolytic activities (namely cellulase, xylanase, β-glucosidase, β-xylosidase and α-arabinosidase) required to hydrolyze cellulose and hemicellulose into fermentable sugars. CS ~~includes~~ includes the leaves, stalks, and cobs of the corn plant, and because of its high content of carbohydrates (about 37% cellulose, 20% xylan and 21% lignin by weight) represents a sustainable feedstock for bioethanol production [36]. BSG is the main by-product deriving from breweries after wort manufacture. Produced in large quantities, and mainly used as animal feed, if submitted to an hydrolysis process under adequate conditions, BSG may produce a liquor rich in fermentable sugars that can be used as a source for sweeteners and ethanol production. The selected biomass has been subjected to enzymatic hydrolysis at high temperature immediatly or after pretreatment to compare the efficiency of the diverse enzyme saccharification processes.

Materials and Methods

Materials

ENEA CR Trisaia—Biomass Section, Rotondella (MT), Italy, kindly supplied CS after steam explosion pretreatment at 210°C. Untreated BSG were provided by MALTOVIVO micro-brewery, Avellino, Italy. Glucose, xylose, arabinose, Remazol Brilliant Blue R-D-xylan (RBB-xylan), *p*-nitrophenyl-β-D-xylopyranoside (*p*NPX), *p*-nitrophenyl-β-D-glucopyranoside (*p*NPG) and *p*-nitrophenyl-α-L-arabinofuranoside (*p*NPA) were obtained from Sigma Chemical Co. (St. Louis, MO). AZO-carboxymethylcellulose (AZO-CMC) was purchased from Megazyme International Ireland Ltd. All other reagents were analytical grade.

Sugar Analysis of Raw BSG

BSG was dried at 80°C, milled by a homogenizer until obtaining a fine powder, and stored in sealed vials at room temperature. Analysis of glucose, xylose and arabinose content was performed according to Cruz *et al.* [37]. Briefly, milled BSG was subjected to acid hydrolysis in two steps: the first was performed with 72% sulphuric acid at 30°C for 1 h, and the second with 4% sulphuric acid at 121°C for 1 h after dilution of the medium. The resulting hydrolysate was analysed for glucose, xylose and arabinose content by HPLC as described later.

BSG Pretreatment

A number of BSG pretreatments were carried out in order to choose those more functional for the next phase of enzyme saccharification. Milled BSG was suspended in distilled water (1% w/v) and treated at 700 Watt and 2.4 GHz in a commercial microwave oven for 1 min, or suspended in 1% sulphuric acid and heated at 60°C for 3 h. Alternatively, BSG was suspended in 0.45% sulphuric acid, and then heated at 121°C for 15 min [38]. This last pretreatment was carried out by increasing the amount of BSG (1, 5 and 10% w/v).

Enzyme Cocktail Preparation and Enzyme Assays

Glycolytic enzymes required for biomass saccharification were extracted from the cell membranes of *S. Solfataricus*, strain Oα as follows: wet cells were suspended in 50 mM Tris-HCl buffer added with 0.7 mM PMSF (pH 7.0) and broken by ultrasonic cell disruptor (Sonifier Braun Labsonic). The crude extract was centrifuged at 2,000 x *g* for 10 min at 4°C, and the opalescent supernatant was cleared by ultracentrifugation at 50,000 x *g* for 1 h at 4°C. The pellet, constituted by membrane fragments, was suspended in 50 mM Tris-HCl buffer added with 0.5% (v/v) Triton X-100 (pH 7.0), and incubated at 70°C for 18 h. The sample was ultracentrifuged at the same conditions above described, and the supernatant containing the enzyme activities of interest (named enzyme cocktail), was dialysed against 50 mM sodium acetate buffer (pH 5.0).

Activities in the enzyme cocktail were measured at 80°C and pH 5.0 as follows: cellulase activity was determined with the soluble chromogenic substrate AZO-CMC. The standard assay mixture, containing 250 µl 0.2% AZO-CMC in 50 mM sodium acetate buffer, and 250 µl enzyme plus buffer, was incubated at the chosen temperature for 30 min. The reaction was stopped by adding 1.25 ml of 96% ethanol to the mixture, followed by incubation at room temperature for 10 min and centrifugation at 16,100 x *g* for 10 min. The absorbance of the supernatant was measured at 590 nm, and the enzyme units were calculated from a standard curve constructed with known amounts of cellulase from *Trichoderma viride*, according to the manufacturer's recommendations. The assay for the xylanase was based on the use of a soluble chromogenic xylan [39]. The activity was measured by adding 250 µl 0.2% RBB-xylan in 50 mM sodium acetate buffer to 100 µl of enzyme solution, and incubating for 30 min. The reaction was stopped by addition of 1 ml of 96% ethanol to the mixture, followed by incubation at room temperature for 15 min and centrifugation at 16,100 x *g* for 5 min. The absorbance of the supernatant was measured at 590 nm. One unit of xylanase activity was defined as the amount of enzyme required to increase the absorbance at 590 nm of 1 ΔA/min under standard conditions.

β-xylosidase, β-glucosidase and α-arabinosidase activities were measured by using *p*NPX, *p*NPG, and *p*NPA as the substrates, respectively. The appropriately diluted sample was added to 450 µl of substrate (1.5 mM for *p*NPX and *p*NPA, and 2 mM for *p*NPG in 50 mM sodium acetate buffer). The reaction mixture (final volume: 500 µl) was incubated for a variable time ranging from 2 to 5 min and stopped by addition of 1 M Na_2CO_3 (1 ml). The amount of released *p*-nitrophenol was measured at 405 nm. One enzyme unit was defined as the amount of enzyme releasing 1 µmol of *p*-nitrophenol per minute under the described conditions.

Saccharification Process

BSG and CS were separately used to realize a bioconversion process in order to obtain fermentable sugars by using the enzyme cocktail from *S. solfataricus* Oα. The saccharification processes were performed as follows: untreated (1% w/v) and pretreated (1, 5 and 10% w/v) BSG, and steam explosed CS (1% w/v) were mixed with the enzyme cocktail, and incubated at 80°C and pH 4.0 (50 mM sodium acetate buffer), 5.0 (50 mM sodium acetate buffer) or 6.0 (50 mM sodium phosphate buffer) for 24 or 48 h. Aliquots from the reaction mixtures were withdrawn before enzyme addition (time 0) and in the course of enzymatic hydrolysis, and analysed for sugar content as depicted in the following paragraph. Controls with the appropriate buffers in replacement of the enzyme cocktail were carried out under the same reaction conditions.

Analysis of the Sugars Released

For estimation of the fermentable sugars released, the collected samples were extensively centrifuged, and the cleared supernatant was analysed by a high-performance liquid chromatographic system (Dionex, Sunnyvale, CA, USA), equipped with an anionic exchange column (Carbopac PA-100) and a pulsed electrochemical detector (PED). Glucose, xylose and arabinose were separated with 16 mM sodium hydroxide at a flow rate of 0.25 ml/min, and identified by the respective standards. Fucose was used as internal standard.

RESULTS

The enzyme cocktail from *S. Solfataricus* strain Oα contained the main activities required to degrade cellulose and hemicellulose into monosaccharides (Table 1). Biotechnological potentiality of the β-glucosidase was tested by a continuous cellobiose hydrolysis bioreactor [40], and the efficiency of the action the bi-functional β-xylosidase/α-arabinosidase in producing xylose from xylan in cooperation with a xylanase from the thermophilic bacterium *Anoxybacillus flavithermus* was verified in a previous work [41]. As a result, a combination of different BSG ~~preeatments~~ pretreatments followed by saccharification at 80°C with the thermophilic enzyme cocktail, were developed in order to convert cellulose and xylan into fermentable sugars.

Table 1. Glycosyl hydrolases from *Sulfolobus solfataricus* strain Oα contained in the enzyme cocktail used for BSG and CS saccharification

Enzyme	Activity (mU/ml of enzyme cocktail)	Optimal temperature	Optimal pH
Cellulase	29	95	3.5
Xylanase	63	90	4.0
β-Xylosidase	609	85	6.0

α-Arabinosidase	4360	80	6.5
β-Glucosidase	158	> 80	6.5

Figure 1 shows glucose and xylose yields after enzymatic saccharification of two differently pretreated BSG samples in comparison with untreated BSG.

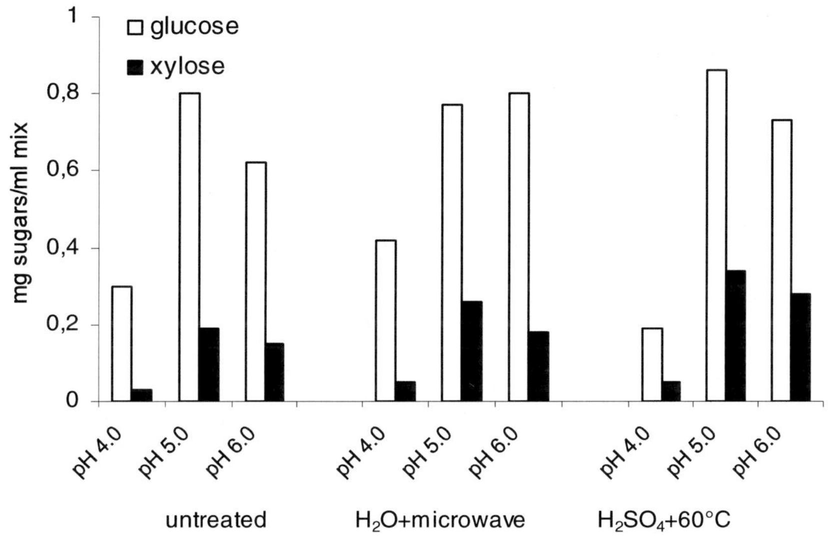

Figure 1. Effect of two diverse pretreatments on glucose and xylose release from BSG. Raw material (1% w/v) was differently pretreated, and subsequently subjected to enzyme saccharification at 80°C for 48 h at various pH values.

Legend of Figure 1 should be justified

Since the enzymes acting on the polysaccharides contained in BSG did not exhibited the same value of optimal pH, the saccharification was performed at three different acidic pH with the aim to ascertain the optimal one for the highest sugar production. The maximum amount of sugars was obtained at pH 5.0 from BSG pretreated with sulphuric acid and thermal cycle at 60°C. However, the amounts of glucose and xylose released were low in comparison to the amount expected for near-total hydrolysis of the polysaccharides (19 and 12.5% of the total estimated glucose and xylose, respectively). Moreover, as only small differences were detected among the untreated and pretreated samples, we attempted a different pretreatment consisting of an hydrolysis of the raw material with more dilute acid, but at higher temperature (121°C). In addition, the BSG concentration was increased with the aim of obtaining an hydrolysate with high content of fermentable sugars.

Very promising results were obtained with this latest pretreatment. The arabinose release was detected in the course of the hydrolysis, contrarily to the previous experiments, and the total amount of sugars released from 1% BSG was 4.5-fold higher than that obtained from 1% BSG pretreated with 1% sulphuric acid at 60°C, confirming the importance of the high temperature in destroying the compactness of the lignocellulosic fibers for an easier action of the enzymes. After 24 h incubation, the total sugars released from the saccharification of 5%

pretreated BSG ~~was~~ were 5-fold higher than the quantity liberated from the saccharification of 1% BSG, proving the good efficiency of the enzymes at these high BSG concentrations (Figure 2). Extension of the saccharification time up to 48 h did not produce further release of sugars.

Additional increase of BSG (10% w/v) did not release supplementary sugars other than those coming from the pretreatment (data not shown). An alternative attempt was carried by adding a 3-fold concentrated enzyme cockatil to the pretreated 10% BSG. In this case, a release of sugars was measured but lower than that expected. Controls were run together to the enzyme samples in order to assure that the release of fermentable sugars was not due to autohydrolysis phenomena of the polysaccharides caused by high temperature and low pH. The monosaccharide composition of the hydrolysate from 5% pretreated BSG was analysed using the high pressure anionic exchange chromatography (Figure 3).

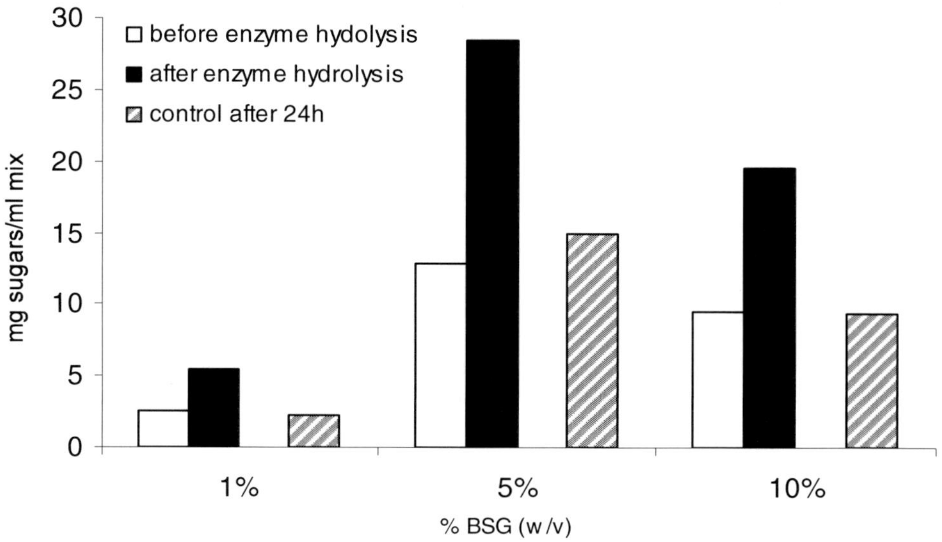

Figure 2. Sugars released from different concentrations of pretreated BSG. BSG was pretreated with 0.45% H_2SO_4 at 121°C for 15 min, and then subjected to saccharification with the enzyme cocktail at 80°C and pH 5.0 for 24 h. Saccharification of 10% BSG was performed with 3-fold concentrated enzyme cockatil. Sugar content was estimated before and after hydrolysis in samples and relative controls.

Legend of Figure 2 should be justified

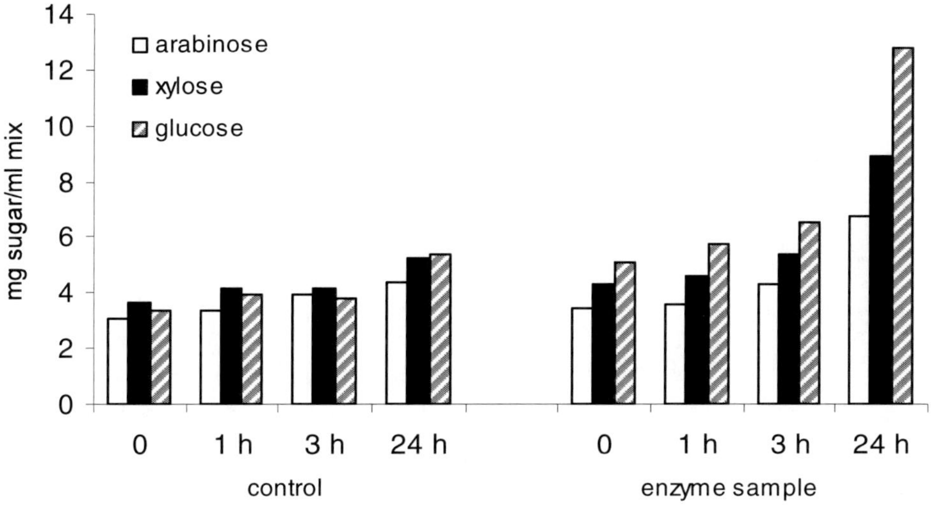

Figure 3. Sugar composition of hydrolysate from 5% BSG pretreated with 0.45% H2SO4 at 121°C for 15 min, and then subjected to enzyme saccharification. Control and sample were incubated for 24 h at 80°C and pH 5.0. Arabinose, glucose and xylose were measured at established times as described in the Materials and methods section.

Legend of Figure 3 should be justified

After 24 h, glucose in the enzyme sample was more than doubled in comparison to the control, proving that sugar production was due to the enzyme action. A smaller difference was observed for xylose and arabinose. However, the quantity of arabinose produced corresponded to that expected after its near-complete release from the feedstock (6.7 mg/ml fermentable mixture), while glucose and xylose in the fermentable mixture represented 64% of their theoretical full release.

Steam-exploded CS (1% w/v) was hydrolyzed with the glycosyl hydrolases from *S. solfataricus* strain Oα at high temperature and different pH values in order to investigate the efficiency of the enzyme cocktail towards another variety of feedstock. The final hydrolysis mixture, qualitatively analyzed by high pressure anionic exchange chromatography, revealed the presence of glucose and xylose as smallest products. Saccharification process carried out at pH 4.0 (Figure 4A) produced lower quantities of sugars in comparison to pH 5.0 (Figure 4B), which resulted the best pH value, as for the BSG saccharification process.

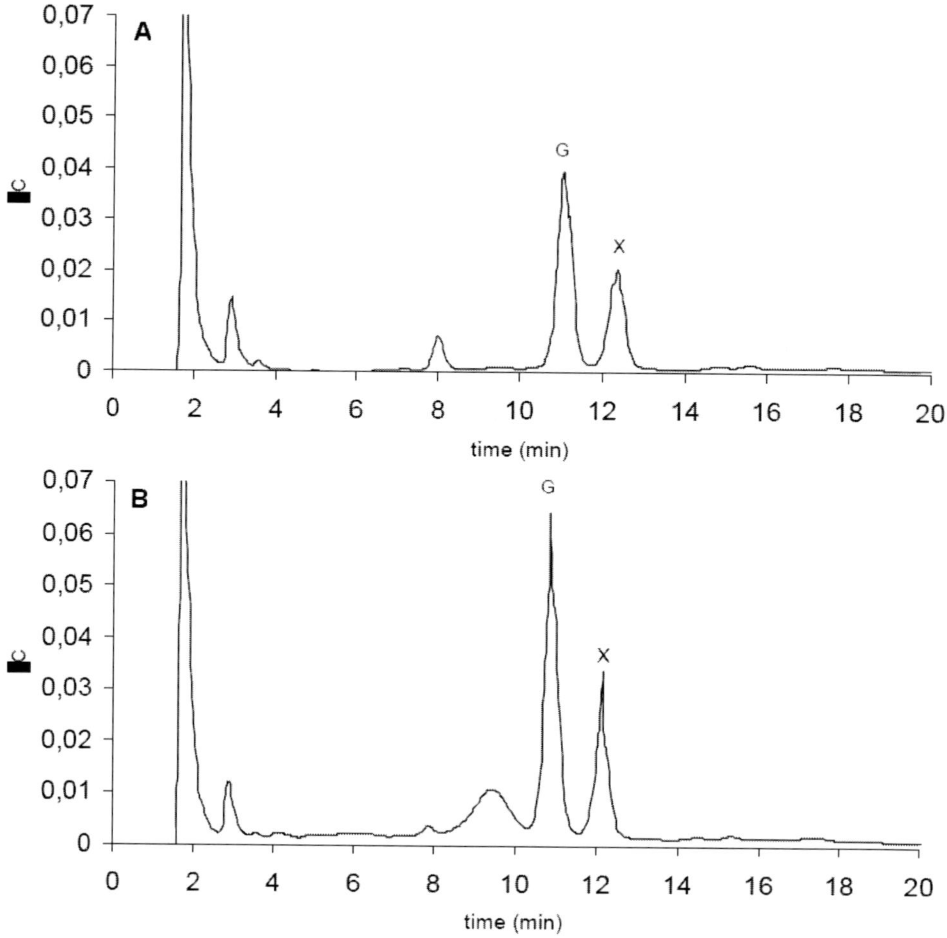

Figure 4. Chromatographic profile of monosaccharides released after 24 h at 80°C from steam-explosed CS at pH 4.0 (A) and 5.0 (B). Glucose (G), xylose (X). Chromatographic conditions are described in the Materials and methods section.

Legend of Figure 4 should be justified; Y Axis should be μC

Conclusion

Thermostable enzymes offer potential benefits in the hydrolysis of lignocellulosic substrates. In fact, their enhanced stability to high temperatures, low and high pH values and organic solvents, allows improved hydrolysis performance and increased flexibility with respect to process configurations, leading to improvement of the overall economy of the process [42]. The successful hydrolysis of lignocellulosic biomass to fermentable sugars with these thermostable enzymes could provide an inexpensive sugar source and could have an enormous economic potential for the conversion of plant biomass into bioethanol. In this context, several attractive glycosyl hydrolase activities were extracted from cell-membranes of the hyperthermophilic archaeon *S. solfataricus* Oα and utilized in experiment sets of saccharification of untreated and pretreated brewer's spent grains and corn stover. Although lignocellulosic materials represent the most abundant polysaccharides in nature, they still don't have significant utilization for the difficulties encountered in their hydrolysis. Therefore, suitable pretreatments are a key step if an efficient enzyme hydrolysis for obtaining fermentable sugars is desired. Thermal pretreatments with acid aqueous solutions or saturated steam are reported to break down the solid matrix by partially hydrolyzing chemical bonds [7, 16]. Acid pretreatment is the most widely tested approach, and even if nitric acid [43], hydrochloric acid [44] and phosphoric acid [45] have also been tested, the use of dilute sulphuric acid is the most generally applied [46, 47]. We attempted several pretreatments of BSG in order to make the polysaccharides more available to the thermostable hydrolytic enzymes. Methods based on the use of microwave or dilute acids were investigated, and the best result, in terms of sugar release following enzyme saccharification, was observed after pretreatment of BSG with sulphuric acid at high temperature. The failure of the pretreatment carried out at lower temperature and higher acid concentration was the demonstration that the high temperature represents a significant aspect for the defibering of lignocellulose, even if acting in combination with more dilute acid. Among the pH values chosen for the realization of the saccharification process, pH 5.0 was the best one. As depolymerizing enzymes, namely cellulase and xylanase, show optimum pH at 3.5 and 4.0, respectively, and the remaining glycosidases at pH 6.0 or more, pH 5.0 could represent the correct compromise to which all the activities work together with good synergy. The saccharification process of pretreated BSG was accomplished by increasing the substrate concentrations with the aim to produce a hydrolysis mixture with a very high sugar content for a high yield fermentation step. The amount of released monosaccharides was proportional when hydrolyzing 1 and 5% pretreated BSG, whereas no sugar release, due to enzyme action, was osbserved for all the duration of the 10% BSG saccharification process. This effect could be caused by an elevated concentration of low molecular weight compounds liberated by the pretreatment, that negatively affected the enzyme action (inhibitor or toxic). Further experiments with a more concentrated enzyme cocktail led to the hydrolysis of cellulose and hemicellulose to a certain extent proving that enzymes were still capable of acting on the substrates, but some molecules in the pretrated mixture were responsible for the reduced activity. More analyses are in progress to elucidate the nature of these phenomena.

Enzyme saccharification of 5% pretreated BSG produced high conversion of substrates into fermentable sugars. Glucose yield (64%) was calculated on the basis of the total glucose measured after analysis of the raw material (39.8 g/100 g BSG). The percentage of cellulose conversion should be underestimated because the amount of cellulose in BSG is usually lower (around 17–22%), and the higher content of glucose estimated in our raw material could come not only from cellulose but also from malt residues (Table 2) [48, 49].

Table 2. Polysaccharide contents in brewer's spent grain

g/100 g BSG	Cellulose	Xylan Arabinan
Mussatto[48]	16.8 – 21.9	19.8 – 20.6 8.5 – 9.4
Faulds [49]	17.0	39.0[a]

Cruz [37]	39.8 (as free glucose)	27.5 (as free xylose) 13.2 (as free arabinose)

[a]Reported as the sum of xylan and arabinan.

Third column of Table 2: Xylan Arabinan should be written with more separation. Also the values 19.8 – 20.6 8.5 – 9.4 should be inserted under Xylan and Arabinan, respectively with a better separation. The same for 27.5 (as free xylose) 13.2 (as free arabinose) under Xylan and Arabinan, respectively.

Since the combination of dilute acid with high temperature provided encouraging results, further studies are in progress to select the best operational parameters (e.g. substrate and enzyme cocktail concentrations) in order to maximize the release of the fermentable sugars. However, the major problem of acid hydrolysis is that the decomposition of monomeric sugars produced during the reaction takes place simultaneously with the hydrolysis of polysaccharides: xylose and arabinose can be decomposed into furfural, and glucose into hydroxymethylfurfural (HMF). Other compounds that may often be formed during an acid hydrolysis process include acetic acid and lignin degradation products (LDPs). Furfural, HMF, acetic acid and LDPs are known to be potent inhibitors of yeast growth [50]. For this reason, before performing a successful bioconversion process, it is necessary to establish the most useful operational conditions in order to produce a liquor with few toxic compounds.

Therefore, chemical detoxification of the hydrolysis mixture can be proposed to remove toxic compounds, and, moreover, recovery of the enzymes, that can be easily separated from the unhydrolyzed substrate through centrifugation and from low molecules by dialysis, can be realized with the purpose of recycling the enzyme cocktail for a new saccharification process. Promising results were also obtained from enzyme saccharification of a steam-~~exposed~~ explosed CS. When a sample of CS was hydrolyzed at acidic pH and high temperature by the cocktail of glycosyl hydrolases from *S. solfataricus*, glucose and xylose were released with little differences in relation to the different pH of the reaction media. These preliminary data on saccharification of CS demonstrate the efficacy of the cooperative action shared by the thermostable enzymes, in achieving the degradation of the polysaccharides to fermentable sugars in lignocellulosic biomass from different origin.

In conclusion, thermoacidophilic enzymes appear to be attractive candidates to be utilized in the biofuel industry for fermentable sugar production after thermal pretreatment in an acidic environment of several types of agricultural waste.

REFERENCES

[1] Wheals, AE; Basso, LC; Alves, DMG; Amorim, HV. Fuel ethanol after 25 years. *Trends Biotechnol*, 1999 17, 482-487.

[2] Monique, H; Faaij, A; van den Broek, R; Berndes, G; Gielen, D; Turkenburg, W. Exploration of the ranges of the global potential of biomass for energy. *Biomass and Bioenergy,* 2003 25, 119 -133.

[3] Taherzadeh, MJ; Karimi, K. Acid-based hydrolysis processes for ethanol from lignocellulosic materials: A review. *BioRes*, 2007 2, 472-499.

[4] Zaldivar, J; Nielsen, J; Olsson, L. Fuel ethanol production from lignocellulose: a challenge for metabolic engineering and process integration. *Appl Microbiol Biotechnol*, 2001 56, 17-34.

[5] Chandrakant, P; Bisaria, VS. Simultaneous conversion of cellulose and hemicellulose to ethanol. *Crit Rev Biotechnol*, 1998 18, 4, 295-331.

[6] Gray, KA; Zhao, L; Emptage, M. Bioethanol. *Curr Opin Chem Biol*, 2006 10, 141-146.

[7] Sun, Y; Cheng, JY. Hydrolysis of lignocellulosic materials for ethanol production: a review. *Bioresour Technol*, 2002 83, 1-11.

[8] McMillan, JD. Conversion of hemicellulose hydrolyzates to ethanol. In: (eds) *Enzymatic conversion of biomass for fuels production*, 411-437; 1994.

[9] Taherzadeh, MJ; Karimi, K. Enzymatic-based hydrolysis processes for ethanol from lignocellulosic materials: A review. *BioRes*, 2007 2, 707-738.

[10] Chang, VS; Holtzapple, MT. Fundamental factors affecting biomass enzymatic reactivity. *Appl Biochem Biotechnol*, 2000 84, 25-38.

[11] Kim, S; Holtzapple, MT. Effect of structural features on enzyme digestibility of corn stover. *Bioresour Technol*, 2006 97, 583-591.

[12] McMillan, JD. Pretreatment of lignocellulosic biomass. In: Himmel ME, Baker JO, Overend RP (eds.). *Enzymatic Conversion of Biomass for Fuels Production*. American Chemical Society, Washington, DC, 292–324; 1994.

[13] Chandra, RR; Bura, R; Mabee, WE; Berlin, A; Pan, X; Saddler, JN. Substrate pretreatment: the key to effective enzymatic hydrolysis of lignocellulosics.? *Adv Biochem Eng Biotechnol,* 2007 108, 67-93.

[14] Taherzadeh, MJ; Karimi, K. Pretreatment of lignocellulosic wastes to improve ethanol and biogas production: a review. *Int J Mol Sci*, 2008 9, 1621-1651.

[15] Mosier, N; Wyman, C; Dale, B; Elander, R; Lee, YY; Holtzapple, M; Ladisch, M. Features of promising technologies for pretreatment of lignocellulosic biomass. *Bioresour Technol*, 2005 96, 673-686.

[16] Lasa, Y; Berenguer, J. Thermophilic enzymes and their biotechnological potential. *Microbiologia*, 1993 9, 77-89.

[17] Haki, GD; Rakshit, SK. Developments in industrially important thermostable enzymes: a review. *Bioresour Technol*, 2003 89, 17-34.

[18] Blumer-Schuette, SE; Kataeva, I; Westpheling, J; Adams, MWW; Kelly, RM. Extremely thermophilic microorganisms for biomass conversion: status and prospects. *Curr Opin Biotechnol*, 2008 19, 210-217.

[19] Bergquist, PL; Gibbs, MD; Morris; DD; Teò, VS; Jsaul, DJ; Morgan, HW. Molecular diversity of thermophilic cellulolytic and hemicellulolytic bacteria. *FEMS Microbiol Ecol,* 1999 28, 99-110.

[20] Gilad, R; Rabinovich, L; Yaron, S; Bayer, EA; Lamed, R; Gilbert, HJ; Shoham, Y. CelI, a noncellulosomal family 9 enzyme from *Clostridium thermocellum*, is a processive endoglucanase that degrades crystalline cellulose. *J Bacteriol*, 2003 185, 391-398.

[21] Zverlov, VV; Schantz, N; Schwarz, WH. A major new component in the cellulosome of *Clostridium thermocellum* is a peocessive endo-beta-1,4-glucanase producing cellotetraose. *FEMS Microbiol Lett*, 2005 249, 353-358.

[22] Te'o, VS; Saul, DJ; Bergquist, PL. CelA, another gene coding for a multidomain cellulase from the extreme thermophile *Caldocellum saccharolyticum*. *Appl Microbiol Biotechnol*, 1995 43, 291-296.

[23] Bauer, MW; Driskill, LE; Callen, W; Snead, MA; Mathur, EJ; Kelly; RM. An endoglucanase, EglA, from the hyperthermophilic archaeon *Pyrococcus furiosus* hydrolyzes beta-1,4 bonds in mixed-linkage (1-->3),(1-->4)-beta-D-glucans and cellulose. *J Bacteriol.* 1999 181, 284-290.

[24] Winterhalter, C; Liebl, W. Two extremely thermostable xylanases of the hyperthermophilic bacterium *Thermotoga maritima* MSB8. *Appl Environ Microbio*, 1995 61, 1810-1815.

[25] Liebl, W. Cellulolytic enzymes from Thermotoga species. *Methods Enzymol*, 2001 330, 290-300.

[26] Bronnenmeier, K; Kern, A; Liebl, W; Staudenbauer, WL. Purification of *Thermotoga maritima* enzymes for the degradation of cellulosic materials. *Appl Environ Microbiol*, 1995 61, 1339-1407.

[27] Asha Poorna, C; Prema, P. Production of cellulase-free endoxylanase from novel alkalophilic thermotolerant *Bacillus pumilus* by solid-state fermentation and its application in wastepaper recycling. *Bioresour Technol*, 2007 9, 485-490.

[28] Eckert, K; Zielinski, F; Lo Leggio, L; Schneider, E. Gene cloning, sequencing and characterization of a family 9 endoglucanase (CelA) with an unusual pattern of activity from the thermoacidophile *Alicyclobacillus acidocaldarius* ATCC27009. *Appl Microbiol Biotechnol*, 2002 60, 428-436.

[29] Morana, A; Esposito, A; Maurelli, L; Ruggiero, G; Ionata, E; Rossi, M; La Cara, F. A novel thermoacidophilic cellulase from *Alicyclobacillus acidocaldarius*. *Protein Pept Lett*, 2008 15, 1017-1021.

[30] Huang, Y; Krauss, G; Cottaz, S; Driguez, H; Lipps, G. A highly acid-stable and thermostable endo-beta-glucanase from the thermoacidophilic archaeon Sulfolobus solfataricus. *Biochem J*, 2005 385, 581-588.

[31] Cannio, R; Di Prizito, N; Rossi, M; Morana, A. A xylan-degrading strain of *Sulfolobus solfataricus*: isolation and characterization of the xylanase activity. *Extremophiles*, 2004 8, 117-124.

[32] De Rosa, M; Gambacorta, A; Bu'lock, JD. Extremely thermophilic acidophilic bacteria convergent with *Sulfolobus acidocaldarius*. *J Gen Microbiol*, 1975 86, 156-164.

[33] Maurelli, L; Giovane, A; Esposito, A; Moracci, M; Fiume, I; Rossi, M; Morana, A. Evidence that the xylanase activity from *Sulfolobus solfataricus* Oα is encoded by the endoglucanase precursor gene (*sso1354*) and characterization of the associated cellulase activity. *Extremophiles*, 2008 12, 689-700.

[34] Nucci, R; Moracci, M; Vaccaro, C; Vespa, N; Rossi, M. Exo-glucosidase activity and substrate specificity of the beta-glycosidase isolated from the extreme thermophile *Sulfolobus solfataricus*. *Biotechnol Appl Biochem.* 1993 17, 239-250.

[35] Morana, A; Paris, O; Maurelli, L; Rossi, M; Cannio, R. Gene cloning and expression in *Escherichia coli* of a bi-functional beta-D-xylosidase/alpha-L-arabinosidase from *Sulfolobus solfataricus* involved in xylan degradation. *Extremophiles*, 2007 11, 123-132.

[36] Kadam, KL; McMillan, JD. Availability of corn stover as a sustainable feedstock for bioethanol production. *Bioresour Technol*, 2003 88, 17-25.

[37] Cruz, JM; Moldes, AB; Bustos, G; Torrado, A; Dominguez, JM. Integral utilisation of barley husk for the production of food additives. *J Sci Food Agri.*, 2007 87, 1000-1008.

[38] White, JS; Yohannan, BK; Walker, GM. Bioconversion of brewer's spent grains to bioethanol. *FEMS Yeast Res*, 2008 8, 1175-1184.

[39] Biely, P; Mislovicova, D; Toman, R. Soluble chromogenic substrates for the assay of endo-1,4-beta-xylanases and endo-1,4-beta-glucanases. *Anal Biochem*, 1985 144, 142-146.

[40] D'Auria, S; Pellino, F; La Cara; F; Barone, R; Rossi, M; Nucci, R. Immobilization on chitosan of a thermophilic beta-glycosidase expressed in *Saccharomyces cerevisiae. Appl Biochem Biotechnol*, 1996 61, 157-166.

[41] Kambourova, M; Mandeva, R; Fiume, I; Maurelli, L; Rossi, M; Morana; A. Hydrolysis of xylan at high temperature by co-action of the xylanase from *Anoxybacillus flavithermus* BC and the beta-xylosidase/alpha-arabinosidase from *Sulfolobus solfataricus* Oα. *J Appl Microbiol*, 2007 102, 1586-1593.

[42] Viikari, L; Alapuranen, M; Puranen, T; Vehmaanperä, J; Siika-Aho, M. Thermostable enzymes in lignocellulose hydrolysis. *Adv Biochem Eng Biotechnol,* 2007 108, 121-145.

[43] Brink, D.L. Method of treating biomass material. *US Patent*, 1994 5, 366-558.

[44] Goldstein, IS; Easter, JM. An improved process for converting cellulose to ethanol. *TAPPI Journal,* 1992 75, 135-140.

[45] Israilides, CJ; Grant, GA; Han, YW. Sugar level, fermentability, and acceptability of straw treated with different acids. *Appl Environ Microbiol*, 1978 36, 43-46.

[46] Kim, JS; Lee, YY; Park, SC. Pretreatment of wastepaper and pulp mill sludge by aqueous ammonia and hydrogen peroxide. *Appl Biochem Biotechnol*, 2000 84, 129-139.

[47] Nguyen, QA; Tucker, MP; Keller, FA.; Eddy, FP. Two stage dilute-acid pretreatment of softwoods. *Appl Biochem Biotechnol*, 2000 86, 561-576.

[48] Mussatto, SI; Robert, IC. Chemical characterization and liberation of pentose sugars from brewer's spent grain. *J Chem Technol Biotech*, 2006 81, 268-274.

[49] Faulds, CB; Sancho, AI; Bartolome, B. Mono- and dimeric ferulic acid release from brewer's spent grain by fungal feruloyl esterases. *Appl Microbiol Biotechnol*, 2002 60, 489-493.

[50] Palmqvist, E; Hahn-H¨agerdal, B. Fermentation of lignocellulosic hydrolysates. II: Inhibitors and mechanisms of inhibition. *Bioresour Technol*, 2000 74, 25-33.

In: Bioethanol: Production, Benefits and Economics ISBN 978-1-60741-697-5
Editor: Jason B. Erbaum

Chapter 5

Bioethanol from Starch or Sugar: Energy Security and Life Cycle Environmental Impacts

L. Reijnders*

IBED, University of Amsterdam, Nieuwe Achtergracht 166
1018 WV the Netherlands

Abstract

Life cycle studies of fuel ethanol made from starch or sugar show that such ethanol tends to do not better or worse than conventional gasoline as to the emission of eutrophying and acidifying substances. Life cycle assessments are not unanimous regarding the emission of ecotoxic substances. Fuel ethanol varieties from different crops vary much in their life cycle emissions of substances that contribute to oxidizing smog. Life cycle studies of bioethanol from specified crops show diverging results regarding cumulative fossil fuel demand and the emission of greenhouse gases. If properly done and if the allocation of fuel demand to ethanol and co-products is on the basis of prices, cumulative fossil fuel demand is relatively high for ethanol currently produced from European grain or U.S. corn and relatively low for ethanol from Brazilian sugarcane, making the latter relatively conducive to energy security. The 'seed-to-wheel' emissions of greenhouse gases associated with bioethanol produced from sugar or starch are often higher than the corresponding life cycle emissions of conventional gasoline A major reduction of greenhouse emissions may be possible when cropping for bioethanol production is restricted to abandoned soils which currently sequester little carbon.

1. Introduction

There is a rapid growth in the production and use of bioethanol to displace fossil hydrocarbons in gasoline (Sanchez and Cardona 2008; Macedo et al. 2008; OECD 2008; von

* Tel.: + 31-20-5256206; Fax: + 31-20- 5257431; E-mail: l.reijnders@uva.nl.

Braun 2008). The bioethanol market is currently dominated by bioethanol produced from starch or sugar. Sugarcane, sugarbeet, cassava, corn (maize), rye and wheat are the most important crops used for bioethanol production (Reijnders 2008). 'Second-generation' lignocellulosic ethanol is not expected to offer a large scale alternative to current sugar and starch based ethanol fuels before 2020 (Gibbs et al. 2008; OECD 2008)

Life cycle assessments have been made to investigate how well bioethanol made from sugar or starch does with respect to current gasoline. Life cycle assessments consider bioethanol 'seed-to-wheel', though there are in practice also assessments that consider a part of the complete life cycle (for instance 'seed-to-factory- gate' or 'seed-to-tank'). The life cycle assessments which have been carried out, cover a variety of environmentally relevant aspects of biofuels, such as cumulative energy- and fossil fuel demand, (net) emission of greenhouse gases, and emissions that may contribute to eutrophication, acidification, oxidizing smog and ecotoxicity.

The life cycle assessments published so far agree that bioethanol tends to do no better or worse than conventional fossil fuels in the emission of compounds that contribute to acidification and eutrophication (Sheehan et al. 2004; Reinhardt et al. 2006; Zah et al. 2007; Halleux et al. 2008; Kim and Dale 2008; Nguyen and Gheewala 2008).

Available studies are not unanimous regarding the emission of ecotoxic substances. Zah et al. (2007) conclude that bioethanol generally does better than fossil petrol in this respect, whereas Halleux et al. (2008) suggest that bioethanol from sugarbeet does worse.

Current varieties of bioethanol vary in life cycle emissions of compounds which may contribute to oxidizing smog. In this respect, ethanol from sugarcane tends to do worse than conventional gasoline Zah et al. 2007; Nguyen and Gheewala 2008b), whereas Zah et al (2007) find that bioethanol from wheat or corn does better than gasoline and Halleux et al. (2008) conclude that bioethanol from sugarbeet does worse than gasoline.

The life cycle studies performed so far give remarkably different assessments about two other environmentally relevant aspects of fuel ethanol life cycles: cumulative fossil fuel demand and greenhouse gas emissions. The overall divergence between assessments of 'seed-to wheel' or 'seed-to-tank' greenhouse gas emissions for specific varieties of bioethanol is large. Whereas one paper concludes that a variety of bioethanol is in this respect better than its fossil fuel-based competitor, another one concludes that the same biofuel is doing much worse (e.g. Kim and Dale 2005 versus Seachinger et al. 2008 regarding US corn-based ethanol). This is a problem because whether a bioethanol actually does better or worse than a competing fossil fuel as to the emission of greenhouse gases has considerable importance for decision making about the best way forward in limiting climate change. There are also differences in estimates of the cumulative fossil fuel demand for specific varieties of bioethanol (e.g. Kim and Dale 2005 versus Patzek 2004 again regarding US corn-based ethanol), though divergence tends to be smaller than in case of greenhouse gas emissions. Estimates of cumulative fossil fuel inputs are also of interest, as they are a major determinant of the potential of biofuels to displace fossil fuels, which can be important for energy security. The potential to displace fossil fuels may be expressed as net energy yield (e.g. in terms of lower heating value) per hectare (ha^{-1}), which is obtained by subtracting the cumulative fossil fuel input from the biofuel yield ha^{-1}.

In the following, the origins of divergence regarding the life cycle emission of greenhouse gases and cumulative fossil fuel input will be considered. It will turn out that about a number of the causes for current divergence it may be decided that one way to do the life cycle assessment is better than another, but there are also cases in which this is not, or not yet, possible. To the extent that the best way do life cycle assessment can be established, the consequences thereof will be considered.

Section 2 will consider the reasons why life cycle assessments of bioethanol varieties may disagree about greenhouse gas emissions and cumulative fossil fuel inputs.

Section 3 will discuss preferred ways for doing life cycle assessment.

Section 4 will look at the comparison of conventional fossil gasoline and current bioethanol varieties regarding life cycle greenhouse gas emissions on the basis of proper life cycle assessment.

Section 5 will consider estimates of cumulative fossil fuel demand and the net energy yield ha^{-1} of current bioethanol varieties.

Section 6 will briefly focus on the divergence between cumulative fossil fuel demand and life cycle greenhouse gas emissions.

The final section will summarize the main conclusions from this chapter.

2. Why Do Life Cycle Assessments of Transport Biofuel Life Cycles Disagree about Greenhouse Gas Emissions and Cumulative Fossil Fuel Inputs?

There are several reasons why life cycle assessments may disagree about (net) greenhouse gas emissions and cumulative fossil fuel inputs. Some of these reasons are common to both, whereas others refer specifically to greenhouse gas emissions.

Two reasons for divergence refer to both cumulative fossil fuel demand and greenhouse gas emissions. These are:

1. allocation and
2. the scope of the assessment and the estimates of emissions and fossil fuel inputs used.

Ad 1. Allocation. Often a bioethanol is not the only product derived from a crop: there are also co-products. For instance in producing bioethanol from wheat or corn a common co-product is dried distillers grains, whereas glycerol may also be a co-product (Kim and Dale 2005 and 2008; Reijnders 2008; Robinson et al. 2008). Dried distillers grains may e.g. serve as animal feed. The matter then arises: how to allocate greenhouse gas emissions to fuel ethanol and co-products. In practice this can be done based on prices, based on a physical category (such as weight or energy) and on the basis of substitution. In the latter case the environmental burden of a co-product is established on the basis of another, similar product. Different ways to allocate may lead to different outcomes of life cycle assessment (Eickhout et al. 2008; Reijnders 2008). Occasionally there is disagreement about the question whether there should be allocation to a specific co-product at all (Patzek 2006).

Ad 2. The scope of the assessment and the estimates of emissions and fossil fuel inputs used. Assessments may vary in scope and regarding emission data used in life cycle estimates. For instance, whereas some studies consider only operational inputs in bioethanol production, others also consider inputs in creating infrastructure for production and transport (machines, factory buildings, roads, trucks). Emissions and fossil fuel use may vary considerably between sources such as factories and trucks, and choices have to be made about the way to deal with such variety. Moreover emission data may be missing and in such cases estimates have to be made based on extrapolation.

An important reason for divergent estimates of life cycle greenhouse gas emissions are the kinds of emissions considered in life cycle assessment.

In principle there are three important determinants of (net) greenhouse gas emissions linked to the life cycles of bioethanol. They are:

- Carbonaceous greenhouse gas emissions linked to the cumulative demand for fossil fuels
- N_2O emissions linked to fixed N-inputs (e.g. ammonia-based fertilizer) in, and non-product fixed N outputs (e.g. NO_x emissions) of, the fuel ethanol life cycle. The importance of N_2O is linked to its relatively large greenhouse effect. On a molecule for molecule basis N_2O is about 296 times as potent as CO_2, using a 100 year time horizon (Crutzen et al. 2007).
- (Net) biogenic carbonaceous greenhouse gas emissions following from changes in C sequestration (in ecosystems) linked to the fuel ethanol life cycle. These include changes in C sequestration due to direct effects of cultivating biofuel feedstocks (e.g. loss of soil C due to tillage or due to burning of natural vegetation during clearance for sugarcane cultivation). They also include changes in C sequestration due to indirect effects which imply changes in land use. The indirect effects follow from the relative inelasticity of demand for food and fodder (Searchinger et al. 2008; von Braun 2008). For instance, use of corn for fuel ethanol production leads to extra cultivation of crops elsewhere because demand for starch crops is not much reduced when corn is diverted to fuel ethanol production (Searchinger et al 2008).

Cumulative demand for fossil fuels and associated CO_2 emissions are in practice included in life cycle assessment of bioethanol. Non CO_2 -carbonaceous emissions linked to cumulative fossil fuel demand are however often not included in life cycle assessment.

N_2O emissions linked to the fuel ethanol life cycle are not always included and when they are included, off-field emissions are rather often neglected. When N_2O emissions are included there is much uncertainty as to the quantitative aspects thereof (Crutzen et al. 2007; Reijnders and Huijbregts 2008). Such uncertainty is so far rarely reflected in actual assessments. Net biogenic emissions of carbonaceous greenhouse gases are in practice so far often not included in life cycle assessments of fuel ethanol. However there are exceptions (e.g. Reijnders and Huijbregts 2007; Fargione et al. 2008; Gibbs et al. 2008; Searchinger et al. 2008).

3. Preferred Ways for Doing Life Cycle Assessment

Some of the causes of disagreements between life cycle assessment can not or as yet not be eliminated. The first thereof regards allocation. As pointed out in section 2, there are three different ways to allocate. The first is based on prices of the products and co-products. The idea behind this type of allocation is that prices drive production (Weidema 1993). However this way to allocate has problems. Prices are not constants. For instance, the price of glycerol, a co-product of bioethanol production, decreased by a factor 10 between 2004 and 2006 (Yazdani and Gonzalez 2007). Due to this decrease in price, the greenhouse gas emissions allocated to fuel ethanol increases, when allocation is based on prices. A second problem with this way to allocate is that biofuel production is often not driven by its price but by government policy (OECD 2008), which is at variance with the rationale given for allocation on the basis of price (Weidema 1993). The second way to allocate is on a physical basis, e.g. energy. In this case there are curious consequences too. For instance, in this case it is possible to lower the greenhouse gas emission linked to a biofuel strongly by producing large amounts of a low-value co-product. The third way to allocate is based on substitution. For instance, Kim and Dale (2005) have argued that dried distillers grains (a co-product of bioethanol production) may be a substitute for soybean meal in cattle feed, and used the environmental burden of soybean meal to establish the environmental burden of dried distillers grains. However, soybean meal is also a co-product (of soybean oil). And this suggests that substitution of the environmental burden of dried distillers grains by the environmental burden of soybean meal plugs one hole with another. As it stands, the different ways to allocate all have their weak points, and it is not clear that one way is definitely superior. So, it would seem that we will have to live with divergences in outcomes of life cycle assessments linked to different ways to allocate. And this will mean that substantial differences in the outcomes of assessments may remain. This makes it all the more important to clearly state which type of allocation has been used in life cycle assessment.

A second reason for divergence in life cycle assessments of greenhouse gas emissions is linked to the differences in dealing with N_2O emissions (Reijnders 2008). Estimating such emissions is the subject of a lively debate. Moreover, there may be large differences dependent on local conditions. Information about the actual emissions of N_2O from soils relevant to the life cycles of biofuels is as yet largely lacking. So this is a potential source of divergent estimates of life cycle emissions that as yet can not be eliminated. Thus, in the absence of direct measurements it may be best to use a rather wide range of 1.5%-5% for the conversion of fixed N linked to the fuel ethanol life cycle into N_2O (Reijnders and Huijbregts 2008). Such a wide range leads to substantial uncertainty in the outcome of life cycle assessments.

Regarding the coverage of greenhouse gases there is a clear preference: it should be as complete as possible. Fossil fuel linked carbonaceous greenhouse gas emissions, N_2O emission and net biogenic carbonaceous emissions linked to changes in C sequestration should preferentially all be covered. In the latter case the indirect effects of expanding fuel ethanol production caused by the relative inelasticity of demand for food and fodder should also be included. As yet, there are very few life cycle assessments that provide for such comprehensive coverage. Difference in coverage of the biogenic carbonaceous greenhouse gases is probably the major cause of divergence between life cycle assessments of greenhouse gas emissions linked to the life cycles of bioethanol varieties (Reijnders and Huijbregts 2007; Danielsen et al. 2008; Fargione et al. 2008; Gibbs et al. 2008; Searchinger et al. 2008). Also it may be noted that there is considerable uncertainty about the actual impact of growing feedstocks for biofuel production on biogenic C stocks. Estimates from different authors for the same biofuel suggest that the impact of this uncertainty on overall life cycle greenhouse gas emissions may well be in the order of +/-20-40% (Danielsen et al. 2008; Fargione et al. 2008; Reijnders and Huijbregts 2008).

Additionally divergence in estimates of life cycle greenhouse gas emissions may be linked to assumptions that have to be made about the period over which initial changes in C sequestration are to be distributed. The Intergovernmental Panel on Climate Change (IPCC, 2006) has proposed a 20 year time span for such distribution but other time spans – down to 10 years and up to 100 years - are also current (Reijnders and Huijbregts 2008; Wicke et al. 2008). Such differences in time span for the distribution of emissions have a large impact on estimates of life cycle emissions of greenhouse gases. Unfortunately, in the absence of data about the time that arable land converted to fuel ethanol production actually remains in use, it is not possible to decide about a preferable time span for the distribution of carbonaceous emissions due to land use change.

An alternative way to account for change in C stocks is to calculate the 'carbon debt', which can be 'paid back' over a number of years by, if compared with fossil fuels, net lower 'seed-to-wheel' greenhouse gas emissions linked to biofuels from cultivated crops (Danielsen et al. 2008; Fargione et al. 2008; Gibbs et al. 2008).

As to fossil fuel inputs in the fuel ethanol life cycle, coverage of the complete life cycle and of both operational inputs and infrastructure is to be preferred. This is necessary to properly establish cumulative fossil fuel demand of bioethanol. However, it would seem that most of the difference in fossil fuel inputs between fuel ethanol and fossil fuels can be explained by the part of the life cycle ending at the factories that produce those fuels. Also when comparisons are made between fossil fuels and biofuels, it would seem that the major differences in fossil fuel inputs are linked to differences in operational inputs.

4. Greenhouse Gas Emissions of Fuel Ethanol and Conventional Gasoline

When greenhouse gas emissions are covered comprehensively, as outlined in section 3, fuel ethanol from Brazilian sugarcane would seem the best performing variety of bioethanol, provided that the direct and indirect effects of sugarcane cropping on land use do lead to clearing of Cerrado, a savannah (Fargione et al. 2008). In this case the life cycle greenhouse gas emissions of fuel ethanol are somewhat lower than those of conventional gasoline, when the impact of land use change is distributed over a 20 year period as recommended by the IPCC (2006). This would not be the case when sugarcane cropping leads to clearance of rainforest. In the latter case ethanol from sugarcane would do worse than gasoline (Gibbs et al. 2008). Nguyen and Gheewala (2008b) have shown in a study of ethanol made from sugarcane in Thailand, use of coal for providing power and high levels of conversion of sugarcane biomass to methane may lead to this biofuel performing worse than conventional gasoline.

Wheat- or sugarbeet-based bioethanol, the main varieties of fuel ethanol from European soils, do worse as to net greenhouse gas emissions than conventional gasoline if their production expands rapidly, whatever the allocation used (Reijnders and Huijbregts 2007; Gibbs et al. 2008; Soimakallio et al. 2009). The same holds for the most important biofuel in the USA: corn-based ethanol (Searchinger et al. 2008).

The conclusion that fuel ethanol does worse than conventional gasoline regarding greenhouse gas emissions is at variance with many of the estimates of life cycle greenhouse gas emissions of bioethanol which have so far been published (many of them reviewed by von Blottnitz and Curran 2007, also: Halleux et al. 2008; Nguyen and Gheewala 2008a; Quintero et al. 2008). The latter life cycle studies were often more optimistic about fuel ethanol performance. The main reasons for such major differences in outcomes of life cycle assessments would seem that the latter studies do not include the impacts of biofuel production on carbon sequestration in ecosystems due to land use change and land use and that N_2O emissions are not or only partly included.

The quantitatively most important contribution to reducing greenhouse gas emissions linked to cropping for fuel ethanol production may come from restricting such cropping to abandoned soils that currently sequester little carbon (Reijnders and Huijbregts 2009).

As to 'second generation' bioethanol produced from lignocellulosic materials: a rather limited reduction of greenhouse gas emissions linked the fuel ethanol life cycle is possible by converting plant residues to ethanol. The limited character of this reduction largely follows from the need to add large amounts of residues to arable soils in order to maintain soil organic carbon levels (Reijnders and Huijbregts 2009). Organic carbon limits erosion and is conducive to soil fertility. Substituting food crops by lignocellulosic crops for the production of fuel ethanol does not necessarily lead to reduced life cycle emissions of greenhouse gases (Reijnders and Huijbregts 2009).

5. Cumulative Fossil Fuel Demand and Net Energy Yields of Bioethanol

Most studies agree that the 'seed-to-wheel' cumulative demand for fossil fuels associated with fuel ethanol from terrestrial plants is lower than the 'well-to-wheel' demand of (an energetically equivalent amount of) fossil transport fuels (von Blottnitz and Curran 2007; Halleux et al. 2008; Nguyen and Gheewala 2008 aandb; Quintero et al. 2008). However, Patzek and Pimentel (Pimentel 2003; Patzek 2004; Pimentel and Patzek 2005; Patzek 2006) have presented calculations for cornstarch derived ethanol which suggest a higher cumulative fossil fuel demand for bioethanol than for fossil gasoline. The difference in outcome from other studies is partly caused by differences in allocation, partly by higher estimates of fossil fuel input in specified yields from agriculture and in industrial processing, and partly by differences in taking account of the energy demand of the infrastructure (machinery, equipment), labour and seed production needed for fuel ethanol production (Pimentel and Pimentel 2008; Reijnders and Huijbregts 2009).

However, it would seem quite safe to state that in western industrialized countries the cumulative fossil energy demand for bioethanol made from starch, and sugar crops may be quite high, when allocation is on the basis of prices. For ethanol from US corn or European wheat or rye, it would seem unlikely that, when allocated on this basis, the 'seed-to-wheel' cumulative fossil energy demand would be lower than 80% of the energy content of ethanol (Hill et al. 2006; Hammerschlag 2006; Reijnders and Huijbregts 2007; Zah et al. 2007). Cumulative fossil energy demand may be considerably lower for ethanol from sugarcane, especially when lignocellulosic biomass is used for powering processing facilities (Macedo et al. 2008; Almeida D'Agosto and Ribeiro 2009). When the latter applies, for instance the cumulative fossil fuel demand may be lower than 10% of the bioethanol output (Macedo et al.

2008). When allocation is based on the energy content or weight of outputs, cumulative fossil energy demand allocated to fuel ethanol will tend to be lower than in the case of allocation based on prices (Reijnders 2008).

Net energy yields ha^{-1}, which for bioethanol are a measure of the potential to displace fossil fuels, may be calculated by subtracting the lower heating value of cumulative fossil fuel input form the lower heating value of the fuel ethanol produced from a hectare of land.

Data indicative of those net energy yields are in Table 1.

Table 1. Net energy yields in Giga Joules (GJ) per hectare for varieties of fuel ethanol

Crop	Location	Product	Net energy yield in GJ ha^{-1} $year^{-1}$	Source
Sugarcane	Brazil	Ethanol	160-175	Macedo et al. 2008
Starch crops (USA, Europe), sugarbeet (Japan)	USA, Europe, Japan	Ethanol	25-50	Hill et al. 2006; Zah et al. 2007 Koga 2008

From Table 1 it can be concluded that for a specified amount of arable land available, bioethanol from sugarcane is more conducive to achieving independence from imported fossil fuels, or energy security, than ethanol from US corn, Japanese sugarbeet or European starch crops.

6. Divergence between Cumulative Fossil Fuel Demand and Greenhouse Gas Emissions

A remarkable conclusion which can be drawn from this chapter is that 'seed-to-wheel' greenhouse gas emissions may be much at variance with cumulative fossil fuel demand. Bioethanol from sugarcane plantations for which rainforest has been cleared does relatively well in its potential to displace fossil fuels but as pointed out in section 4, such ethanol tends to do worse than fossil gasoline in its emission of greenhouse gases (also Gibbs et al. 2008). Ethanol currently produced from US corn or European wheat has a net energy yield (see Table 1), but does worse than fossil gasoline when all greenhouse gas emissions are taken into account (Gibbs et al. 2008; Searchinger et al. 2008). So, in displacing fossil fuels bioethanol may contribute to energy security but can be counterproductive in limiting climate change. The divergence between fossil fuel displacement and greenhouse gas emissions is linked to the phenomenon that fuel ethanol life cycles do also give rise to emissions of N_2O and to carbonaceous emissions linked to changes in ecosystem C.

Conclusion

Life cycle studies of fuel ethanol made from starch or sugar show that such ethanol does worse or not better than conventional petrol as to the emission of eutrophying and acidifying substances. These studies are not unanimous regarding the relative emission of ecotoxic substances. Varieties of bioethanol vary much in their life cycle emissions of substances that contribute to oxidizing smog. Life cycle studies of specific varieties of fuel ethanol show diverging results regarding cumulative fossil fuel demand and the emission of greenhouse gases. If properly done and when allocation is on the basis of prices, cumulative fossil fuel demand is relatively high for ethanol currently produced from European grains or U.S. corn and relatively low for ethanol from Brazilian sugarcane, making the latter relatively conducive to energy security. The 'seed-to-wheel' emissions of greenhouse gases associated with bioethanol produced from sugar or starch are often higher than the corresponding life cycle emissions of conventional petrol.

REFERENCES

Almeida D'Agosto M and Ribeiro SK 2009. Assessing total and renewable energy in Brazilian automotive fuels. A life cycle inventory (LCI) approach. *Renewable and Sustainable Energy Reviews* in press.

Crutzen PJ, Mosier AR, Smith KA, Winiwarter W 2007. N_2O release from agro-biofuel production negates global warming reduction by replacing fossil fuels. *Atmospheric Chemistry and Physics Discussions* 7: 11191-1120.

Danielsen F, Beukema H Burgess ND, Parish F, Brühl C, Donald PF et al. 2008. Biofuel plantation on forested land: double jeopardy for biodiversity and climate. *Conservation Biology*DOI: 10-1111/j.1523-1739-2008.01096.

Eickhout B, van den Born GJ, Nootenboom J, van Oorschot M, Ros JPM, van Vuuren DP et al. 2008. *Local and global consequences of the EU renewables directive for biofuels*. Bithoven (the Netherlands): Milieu en Natuur Planbureau.

Fargione J, Hill J, Tilman D, Polasky S, Hawthorne P 2008. Land clearing and the biofuel carbon debt. *Science* 319: 1235-1238.

Gibbs HK, Johnston M, Foley JA, Holloway T, Momfreda C, Ramankutty N et al. 2008. Carbon payback times for crop-based biofuel expansion in the tropics: the effects of changing yield and technology. *Environmental Research Letters* 3: 034001.

Halleux H, Lassaux S, Renzoni R, Germain A 2008. Comparative life cycle assessment of two biofuels; ethanol from sugarbeet and rapeseed methyl ester. *International Journal of LCA* 13: 184-190.

Hammerschlag R 2006. Ethanol's energy return n investment: a survey of the literature 1990-present. *Environmental Science and Technology* 40: 1744-1750.

Hill J, Nelson E, Tilman D, Polasky S, Tiffany D. 2006 Environmental, economic, and energetic costs and benefits of biodiesel and ethanol biofuels. *Proceedings of the National Academy of Sciences USA* 103: 11206-11210.

IPCC (Intergovernmental Panel on Climate Change) 2006. *Guidelines for national greenhouse gas inventories*. Volume 4: Agriculture, Forestry and land use. Geneva: IPCC.

Kim S and Dale BE 2005. Life cycle assessment of various cropping systems utilized for producing biofuels: bioethanol and biodiesel. *Biomass and Bioenergy* 29: 426-439.

Kim S and Dale BE 2008 . Life cycle assessment of fuel ethanol derived from corn via dry milling. *Bioresource Technology* 99: 5250-5260.

Koga, N 2008. An energy balance under a conventional crop rotation system in northern Japan: perspectives on fuel ethanol production from sugar beet. *Agriculture, Ecosystems and Environment* 125: 101-110.

Macedo JC, Seabra JEA, Silva JEAR 2008. Greenhouse gas emissions in the production and use of ethanol from sugarcane in Brazil. *Biomass and Bioenergy* 32: 582-595.

Nguyen TLT and Gheewala SH 2008a. Life cycle assessment of fuel ethanol from cassava in Thailand. *International Journal of Life Cycle Assessment* 13: 147-154.

Nguyen TLT and Gheewala SH 2008b. Life cycle assessment of fuel ethanol from cane molasses in Thailand. *International Journal of Life Cycle Assessment* 13: 301-311.

OECD 2008. Directorate for trade and agriculture. *Economic Assessment of biofuel support policies*. Paris: OECD.

Patzek TW 2004. Thermodynamics of the corn-ethanol biofuel cycle. *Critical Reviews in the Plant Sciences* 23: 519-567.

Patzek TW 2006. A first law thermodynamic analysis of the corn-ethanol cycle. *Natural Resources Research* 15: 255-230.

Pimentel D 2003. Ethanol biofuels: energy balance, economics and environmental impact are negative. *Natural Resources Research* 12: 127-134.

Pimentel D and Patzek TW 2005. Thermodynamics of energy production from biomass. *Critical Reviews of Plant Sciences* 24: 327-364.

Pimentel D and Pimentel M 2008. Corn and cellulosic ethanol cause major problems. *Energies* 1: 35-37.

Quintero JA, Montoya MI, Sanchez OJ. Giraldo OH, Cardona CA 2008. Fuel ethanol production from sugarcane and corn: comparative analysis for a Colombian case. *Energy* 33: 385-399.

Reijnders L and Huijbregts MAJ 2007. Life cycle greenhouse gas emissions, fossil fuel demand and solar energy conversion efficiency in European bioethanol production for automotive purposes. *Journal of Cleaner Production* 15: 1806-1812.

Reijnders L 2008.Transport biofuels – a life cycle assessment approach. *CAB Reviews* 3: 071.

Reijnders L and Huijbregts MAJ 2008. Biogenic greenhouse gas emissions linked to the lifecycles of biodiesel derived from European rapeseed and Brazilian soybeans. *Journal of Cleaner Production* 16: 1943-1948.

Reijnders L and Huijbregts MAJ 2009. *Transport biofuels: a seed to wheel perspective* Springer: London.

Reinhardt G, Gärtner S, Patyk A, Rettenmaier N 2006. *Ökobilanzen zu BTL: eine Ökologische Einschätzung* [Biomass to liquids; an environmental assessment]. Heldelberg (Bundesrepublik Deutschland): Institut für Energie- und Umweltforschung Heidelberg GmbH.

Robinson PH, Karges K, Gibson ML 2008. Nutritional evaluation of four c-product feedstuffs from the motor fuel ethanol distillation industry in the Midwestern USA. *Animal Feed Science and Technology* 146: 345-352.

Sanchez OJ and Cardona CA 2008. Trends in biotechnological production of fuel ethanol from different feedstocks. *Bioresource Technology* 13: 5270-5295.

Searchinger T, Heimlich R, Houghton RA, Dong F, Elobeid A, Fabiosa J et al. 2008. Use of U.S. croplands for biofuels increases greenhouse gases through emissions from land use change. *Science* 319: 1238-1240.

Sheehan J, Aden A, Faustian K, Killian K, Brenner J, Walsh M et al. 2004. Energy and environmental aspects of using corn stover for fuel ethanol. *Journal of Industrial Ecology* 7 (3-4) 257-269.

Soimakallio S, Mäkinen T, Ekholm T, Pahkala K, Paapanen T 2009. Greenhouse gas balances of transportation biofuels, electricity and heat generation in Finland – dealing with the uncertainties. *Energy Policy* 37: 80-90.

von Blottnitz H and Curran MA 2007. A review of assessments conducted in bioethanol as transportation fuel from a net energy, greenhouse gas, and environmental life cycle perspective. *Journal of Cleaner Production* 15: 607-619.

von Braun J 2008. *Food prices, biofuels and climate change.* Washington DC: International Food Policy Research Institute.

Weidema BP 1993. Market aspects in product life cycle inventory methodology. *Journal of Cleaner Production* 1: 161-166.

Wicke B, Domburg V, Junginger M, Faaij A 2008. Different palm oil production systems for energy purposes and their greenhouse gas implications. *Biomass and Bioenergy* 32: 1322-1327.

Yazdani SS and Gonzalez R 2007. Anaerobic fermentation of glycerol: a path to economic viability for the biofuels industry. *Current Opinion in Biotechnology* 18: 213-219.

Zah R, Böni H, Gauch M, Hischler R, Lehman M, Wagner P 2007. *Life cycle assessment of energy products: environmental impact assessment of biofuels.* St Gallen, Switzerland: EMPA.

In: Bioethanol: Production, Benefits and Economics
Editor: Jason B. Erbaum
ISBN 978-1-60741-697-5

Chapter 6

ANALYSIS OF ENERGY CONSUMPTION OF DISTILLATION OPTIONS TO OBTAIN HIGH-PURITY BIOETHANOL

Salvador Hernández[a], Juan Gabriel Segovia-Hernández[a], Mariana del Pilar Santamaría-Rivera[a], Héctor Hernández-Escoto[a], Claudia Gutiérrez-Antonio[b], Abel Briones-Ramírez[c,d] and Rafael Maya-Yescas[e]

[a]Universidad de Guanajuato, Campus Guanajuato,
Departamento de Ingeniería Química, Noria Alta s/n,
Guanajuato, Gto., 36050, Mexico
[b]CIATEQ, A.C., Av. del Retablo 150, Col. Fovissste, 76150,
Querétaro, Querétaro, Mexico
[c]Innovación Integral de Sistemas S.A. de C.V., Limas No. 5 Manzana C,
Fraccionamiento Don Manuel, 76114, Querétaro, Querétaro, Mexico
[d]Instituto Tecnológico de Aguascalientes, Departamento de Ingeniería Química,
Av. Adolfo López Mateos #1801 Ote. Fracc. Bonagens,
Aguascalientes, Ags., 20256, Mexico
[e]Facultad de Ingeniería Química, Universidad Michoacana de San Nicolás de Hidalgo,
Ciudad Universitaria, Col. Felicitas del Río, 58060, Morelia, Michoacán, Mexico

ABSTRACT

Among biofuels, bioethanol has increased in importance in many countries because it can be used directly or mixed with gasoline in combustion engines. The production of bioethanol in a fermentative process usually gives a dilute solution from which the bioethanol must be obtained in a high concentration in order to be used as biofuel. The use of bioethanol mixed with gasoline in combustion engines is associated with fewer emissions of both hydrocarbons and carbon monoxide. The production of high purity bioethanol using extractive distillation sequences with ethylene glycol or a dilute solution of NaCl as entrainers are studied in detail in terms of energy consumption and total annual costs. Conventional and complex distillation sequences are designed optimally in a computational framework implemented in Aspen Plus^{TM} and Matlab^{TM}. The results indicate that complex distillation sequences involving thermal linking can reduce energy consumption over conventional distillation sequences using either ethylene glycol or a dilute solution of NaCl as entrainers. As a result, significant reductions in total annual cost can be obtained in the production of high purity bioethanol. This can position ethanol as a competitive biofuel when compared to gasoline.

INTRODUCTION

A large portion of the world economy has been based on processes highly dependent on petroleum; as a consequence, two significant problems have been created: the reduction of oil reserves, and the increase in gas emissions that has been associated with global warming. Researchers in many areas are working on solutions to mitigate these problems. One option that contributes a partial solution is intensification in the production and use of biofuels, including biodiesel, bioethanol, biomass, etc. Of these options, bioethanol is being currently used in combustion engines without modifications in a mixture of up to 20 %. The use of this mixture is important because it allows improved oxidation of hydrocarbons and, as a result, reduction in both hydrocarbons and carbon monoxide emissions (Quintero et al., [1]).

Currently, most bioethanol production is obtained from sugar cane, and, secondly, from corn, although research efforts are focused on industrial production from lignocellulosic material such as agricultural and forest residues. In the production process, four main steps can be identified: Treatment of the raw material to obtain cellulosic mass, saccharification to obtain sugars from the cellulosic mass, fermentation of the sugars, and recovery of ethanol. Independently from the raw material and/or the process used, the product obtained from the fermentation step is a dilute solution of ethanol in water, from which ethanol is separated and purified to the desired concentration. In addition to the research effort in the saccharification and fermentation process, the separation step must also be viewed as a challenge because of the energy that it requires. Assuming that the fermentation process produces a dilute solution of ethanol in water (10 % in moles of ethanol) that requires treatment in order to obtain high purity ethanol that then can be mixed with gasoline [2,3], the production of high purity ethanol using distillation requires significant quantities of energy and mass separation agents such as ethylene glycol, NaCl, KI, or $CaCl_2$.

Distillation is widely used for the separation of many fluid mixtures, but this separation option presents two disadvantages: its high energy consumption and low thermodynamic efficiency (Hernández et al., [4]). Taking into account that distillation is the most important separation option in many industries, researchers and process engineers have been working in order to improve distillation sequences. Two approaches have been used: heat integration in distillation sequences and thermal linking of distillation columns of the sequence. By using these two techniques, important energy savings can be obtained. For instance, in the case of thermally coupled distillation options, the energy savings obtained are in the range of 30 to 50 % over distillation schemes based on conventional distillation columns for the separation of multicomponent mixtures [5-8]. In the case of ternary mixtures (A,B,C), the two conventional distillation sequences are depicted in Figure 1. Figure 1a is the so-called direct distillation sequence in which the components are removed as overhead products, and the distillation sequence shown in Figure 1b is the indirect distillation sequence that recovers the components as bottoms products. It has been reported by Tedder and Rudd [9] that these conventional distillation options are the most appropriate for the separation of mixtures with high contents of either the lightest or heaviest component respectively.

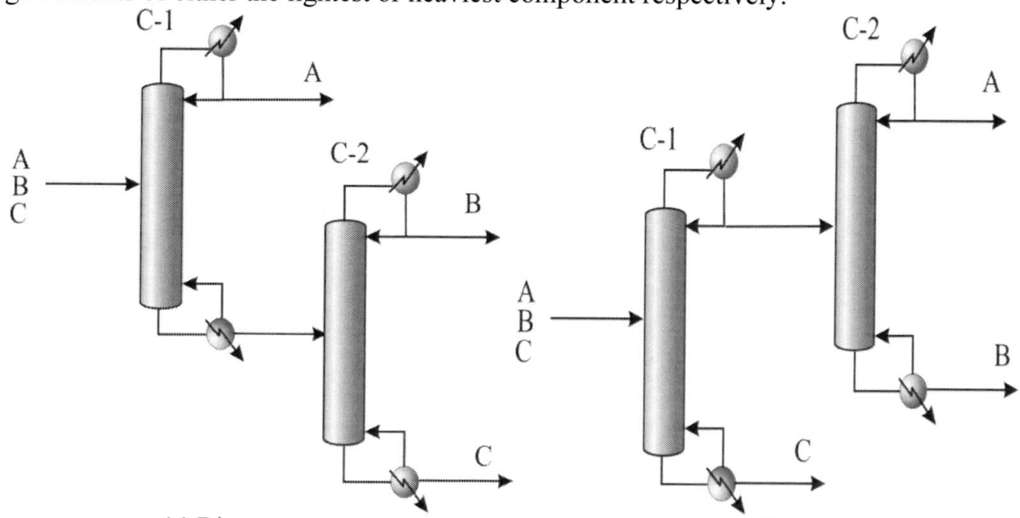

Figure 1. Conventional distillation sequences for ternary separations.

Alternatively, the sequences including thermal links (Figure 2) can be good options for the separation of mixtures with low or high content of the intermediate component. There are many thermally coupled distillation sequences, but those indicated in Figure 2 have been studied in greater detail and some practical implementations have been reported. The thermally coupled distillation sequence indicated in Figure 2a (TCDS-SR) requires a main distillation column coupled to a side rectifier by liquid and vapor streams. This complex distillation option offers energy savings of up to 30% over the conventional direct distillation sequence for the separation of mixtures with less than 15% of the intermediate component in the feed and if the split A/B is easier than B/C. For the same content of the intermediate component, and when the split B/C is easier that A/B, the TCDS-SS (Figure 2b) is the best option. Moreover, the complex distillation option shown in Figure 2c can lower energy consumption up to 50% for the separation of mixtures with high content of the intermediate component in the feed. The last thermally coupled distillation option is called the Petlyuk distillation column; this complex distillation sequence has been implemented in industrial practice by using a dividing wall distillation column. This option uses a single shell divided by a wall (Figure 3), and savings have been reported in both energy and capital costs.

It is important to highlight why the thermally coupled distillation sequences can be better options than conventional distillation sequences for some separations. The explanation can be formulated in terms of the profiles of the intermediate component in the first distillation column of the conventional direct distillation sequence. The composition profile of the intermediate component reaches a maximum below the feed stage and then diminishes as the bottoms part is reached (Figure 4). This effect is known as remixing and is associated with higher energy consumption, because additional energy is required in the second distillation sequence to re-purify the mixture (Triantafyllou and Smith, [10]). A similar effect is observed in the composition profile of the intermediate component in the indirect distillation sequence, but this maximum is reached above the feed stage. This important aspect is taken into account in the design of the dividing wall distillation column, because the side stream is taken from the stage where the composition of the intermediate component presents the maximum.

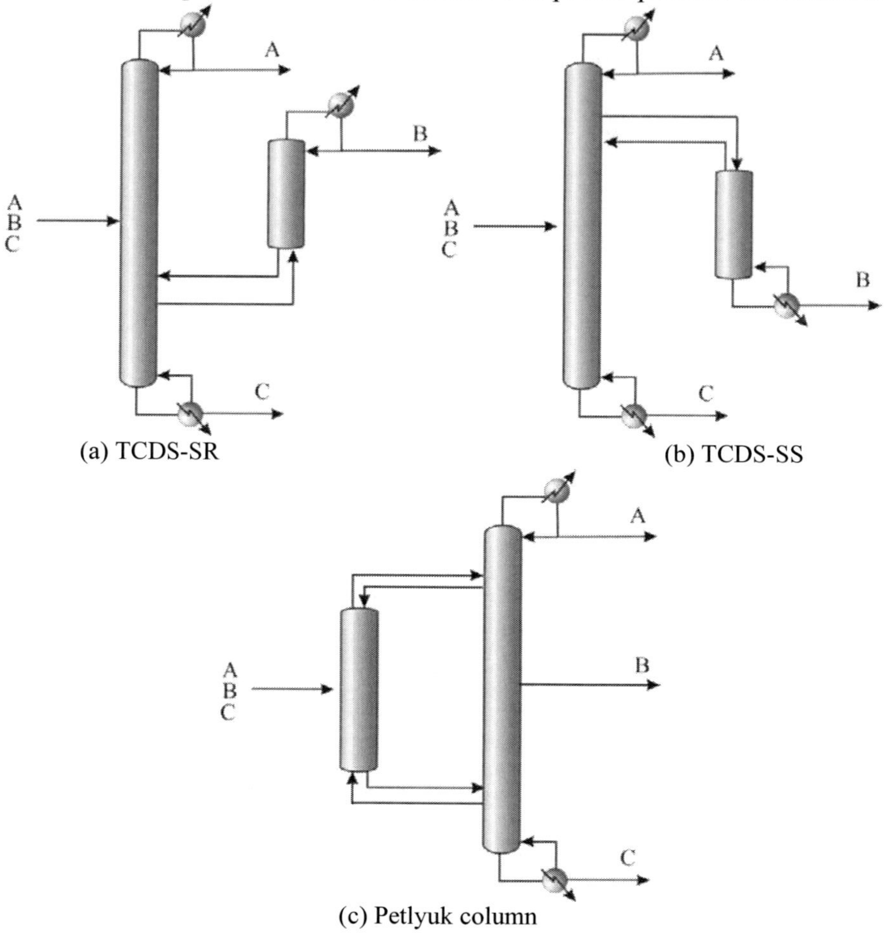

Figure 2. Thermally coupled distillation options for ternary separations: (a) Direct thermally coupled distillation sequence (TCDS-SR), (b) Indirect thermally coupled distillation sequence (TCDS-SS), (c) Petlyuk distillation column.

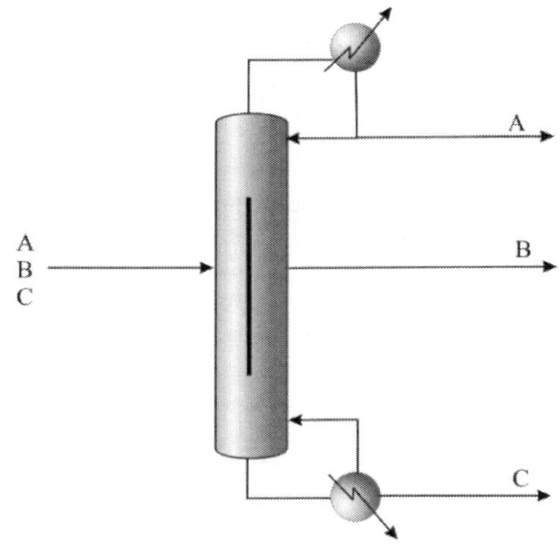

Figure 3. Dividing wall distillation column for ternary separations.

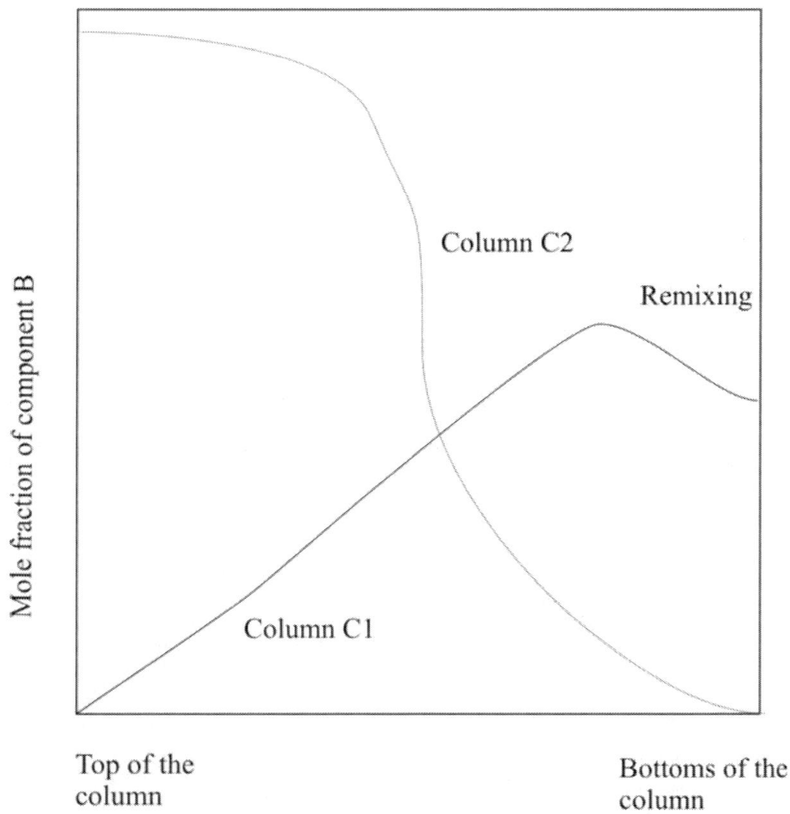

Figure 4. Composition profiles of the intermediate component for the conventional direct distillation sequence.

PROBLEM STATEMENT

As indicated in the preceding section, we are interested in comparing the optimal energy consumption and total annual cost of a several schemes for obtaining high purity ethanol from a dilute solution (10% in moles of ethanol in water). The dilute solution is introduced into a distillation column in order to obtain a distillate product with composition nearly at azeotropic condition (96% in mass of ethanol in water), and almost pure water as bottoms product. The distillate product is introduced into an extractive distillation column using either ethylene glycol or a solution of NaCl as mass separation agents (entrainers). This second stage produces high purity ethanol with a mass fraction of 0.995 and the mass separation agent is recovered. The following distillation options are designed and compared in terms of energy consumption.

The first option is the conventional distillation sequence requiring an additional distillation column using ethylene glycol as entrainer to obtain ethanol as distillate product (mass fraction of 0.995) and a binary mixture of entrainer and water as bottoms product. The bottoms product is introduced into a distillation column to recover the entrainer as shown in Figure 5.

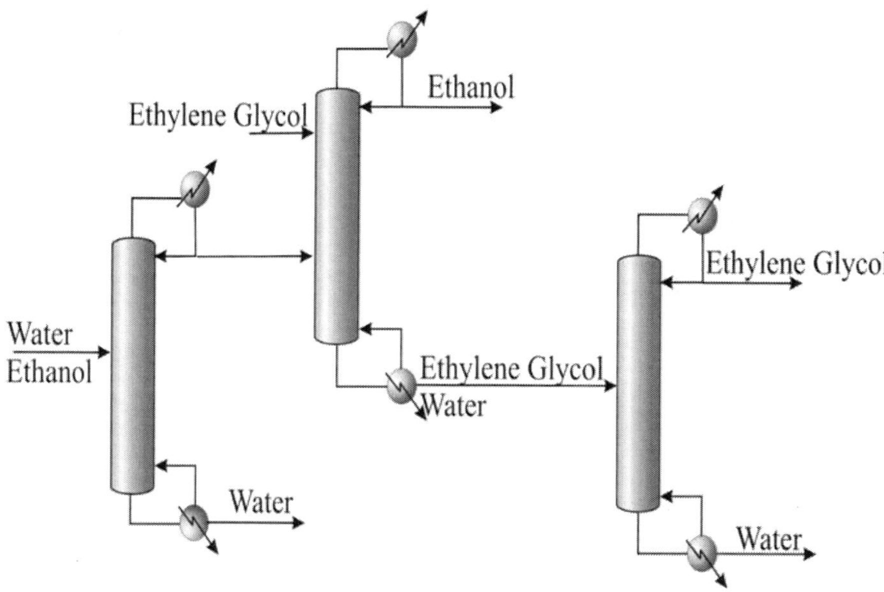

Figure 5. Extractive conventional distillation sequence (ECDS).

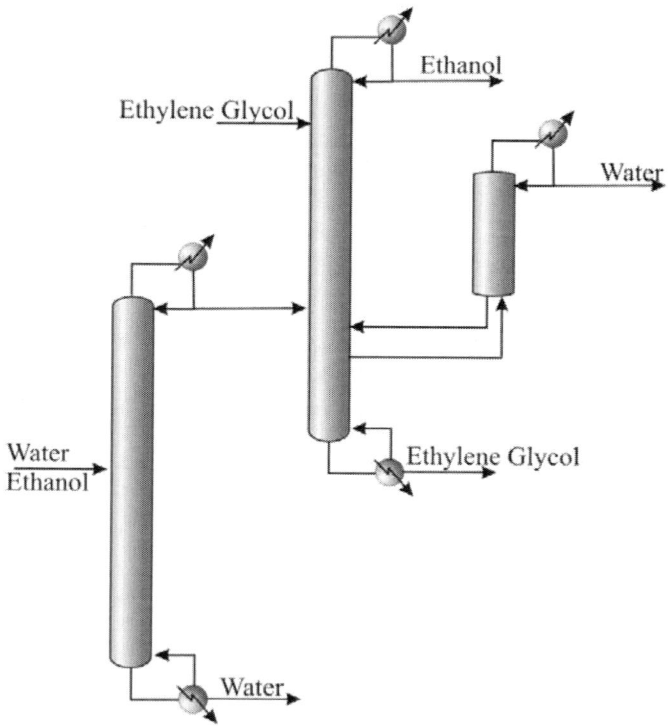

Figure 6. Extractive thermally coupled distillation sequence with a side rectifier (ETCDS-SR).

The second distillation sequence is the thermally coupled distillation sequence with a side rectifier. The distillate of the main column is ethanol with the required purity and the bottoms product of this column is the entrainer. The side rectifier column removes a mixture of ethanol and water that can be recycled to the first distillation column. The distillation option is described in Figure 6; as can be noted, this complex distillation option includes a recycle stream from the side rectifier to the main distillation column. This recycle stream plays an important role in the energy efficiency of the distillation scheme, and it is necessary to detect the value that corresponds to the minimum energy consumption.

The third scheme analyzed involves the use of a distillation column with a side stream. In this case, the distillate product is high purity ethanol, the side stream is a dilute mixture of ethanol and water that can be recycled to the azetropic distillation column and the bottoms product is ethylene glycol (Figure 7).

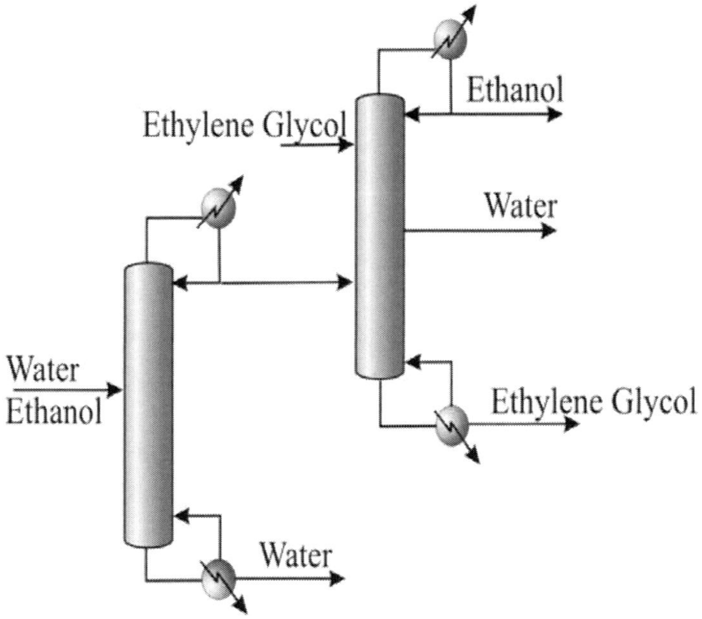

Figure 7. Extractive distillation sequence with a sidestream (EDSS).

The fourth option includes extractive dividing wall distillation sequences (one or two walls). These options recover the ethanol in the distillate and the bottoms product is the entrainer. The sidestream removes a mixture of ethanol and water that can be returned to the first conventional distillation column. These two options are schematically shown in Figures 8 and 9, respectively. These options include two or four recycle streams that must be varied in order to obtain the minimum energy demand in the reboiler. It is important to highlight that these fully thermally coupled distillation sequences can be implemented in industrial practice [11] by using one or two dividing walls as depicted in Figures 8 and 9 respectively.

DESIGN AND OPTIMIZATION PROCEDURES

In order to guarantee minimum energy consumption in the distillation options, an optimization procedure based on genetic algorithms was implemented in Aspen Plus^{TM} and Matlab^{TM}. This computational framework is required because many search variables are involved in the optimization problem. These variables can be continuous or discrete.

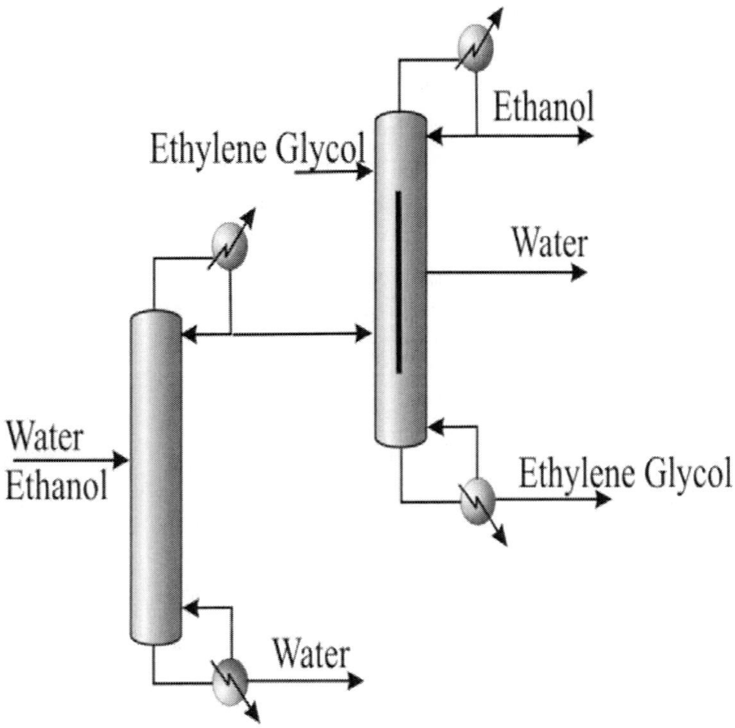

Figure 8. Extractive distillation sequence with one dividing wall (EODWDC).

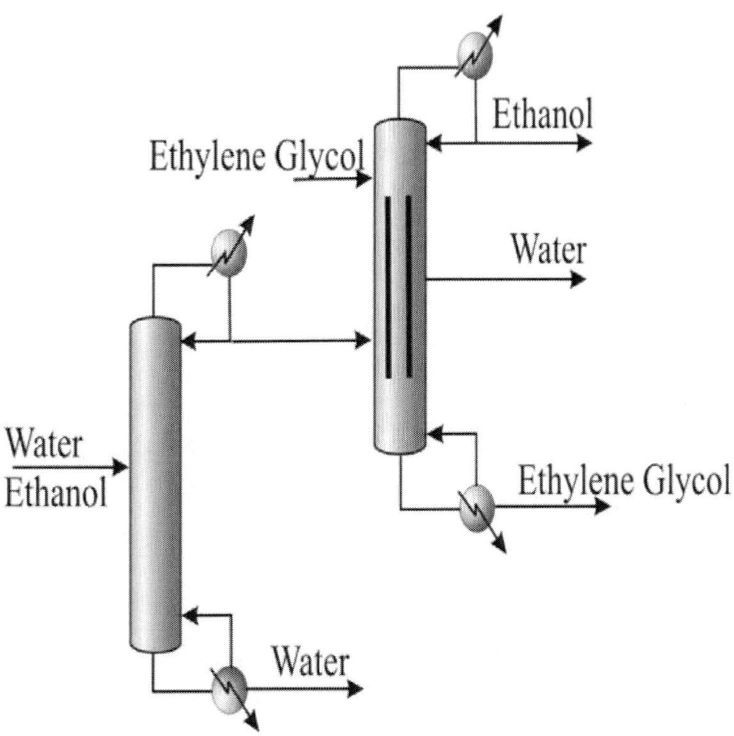

Figure 9. Extractive distillation sequence with two dividing walls (ETDWDC).

The operational pressure was not considered a search variable, since it was set in order to guarantee the use of cooling water in all condensers. All condensers are operated at 14.7 psia using cooling water. Furthermore, it is important to mention that in all cases, the constraint for the composition of ethanol was a mass fraction of 0.995.

The first distillation column used in all the distillation schemes has the minimum number of search variables, but we have continuous and discrete variables in the optimization procedure; for instance, the number of actual stages, feed stage, reflux ratio and distillate rate.

The second distillate column indicated in Figure 5 includes two additional search variables, the feed stage of the entrainer, and the ratio of flows of the entrainer to the feed stream. The procedure for optimal design of the thermally coupled distillation column included in Figure 6 is truly a difficult task, because the recycle stream couples the main distillation column and the side rectifier. This implies that the designs of the columns must be obtained simultaneously. The additional variables in this case are the positions of the liquid and vapor streams between the two columns and the flow of the interconnecting vapor stream.

When Figure 7 is analyzed, we detect important design variables for the second distillation option, for instance, the feed stages, the position and flow of the sidestream.

For the design of the fully coupled distillation options indicated in Figures 8 and 9, the procedure increases the complexity as the number of recycle streams increases; for instance, for one dividing wall there are two recycle streams, meanwhile for two dividing walls, the number of recycle streams increases to four. As we can see, obtaining the optimal designs of the distillation options is a difficult task to complete without the use of a computational framework.

COMPUTATIONAL FRAMEWORK

Obtaining the optimal designs of the extractive distillation sequences is a complex task, since it implies determination of integer and continuous variables. These variables combination must provide the minimal number of stages and heat duty in each column of the sequence. In addition, the purities required in each stream product must be satisfied.

In other words, finding an optimal design means solving a multiobjective, highly nonlinear problem with constraints. The problem is multiobjective because we know beforehand that the number of stages and the heat duty are variables in competition, since we cannot indefinitely decrease one without increasing the other. The problem is highly nonlinear because we are considering the complete set of MESH (component mass balances, equilibrium relationship, summation constraints, energy balance) equations along with the phase equilibrium calculations. Finally, purities and/or recoveries that have to be reached represent the constraints of the problem.

Table 1 presents the number of manipulated variables, along with the number of objectives to be minimized and the constraints considered for each distillation option.

The number of manipulated variables changes because it depends on the complexity of the structure considered; also, the number of reboilers and shells determines the number of objectives that have to be minimized simultaneously. Finally, the purities required are four since we have four product streams; in addition, three recoveries are considered as constraints.

Table 1. Distillation options and their manipulated variables, objectives and constraints

Distillation option	Manipulated variables	Number of objectives		Constraints considered	
	Integer and continuous	Number of stages	Heat duty	Purities	Recoveries
ECDS	14	3	3	4	3
ETCDS-SR	14	3	2	4	3
EDSS	12	2	2	4	3
EODWDC	15	3	2	4	3
ETDWDC	17	4	2	4	3

In order to optimize the conventional and thermal coupled extractive sequences we used the multiobjective genetic algorithm with constraints coupled to Aspen Plus^{TM} developed by Gutiérrez-Antonio et al. [12]. Their algorithm manages the constraints using a multiobjective technique based on the concept of non dominance for constraints-handling proposed by Coello-Coello [13], which guides the NSGA-II search (Meyarivan et al, [14]). Since their code is coupled to commercial simulator Aspen Plus, all the results obtained here considered the rigorous energy and material balances, along with the equilibrium phase calculations.

For all sequences, we used 1600 individuals and 80 generations as parameters of the algorithm. These parameters were obtained through a tuning process, where we performed several runs of the algorithm with different number of individuals and generations.

RESULTS

As indicated above, high purity ethanol will be used pure or mixed with gasoline in combustion engines. For that reason, it is crucial that the composition reach the desired value (mass fraction of ethanol of 0.995) in the extractive distillation columns.

It is important to analyze the composition profiles of the optimal designs obtained in the computational framework. As depicted in Figures 5-9, the role of the first distillation column is the separation of most of the water in the bottoms product; according to Figure 10, the bottoms product is almost pure water obtained from stage number 30. Meanwhile, the top product is an enriched binary mixture of ethanol and water. This stream is sent to the second purification option consisting of extractive distillation schemes. First, the use of the ethylene glycol was analyzed, and the results are presented in Figures 11-13.

Figure 11 presents the composition profiles of the extractive distillation column of Figure 5, where it can be noted that the distillate product is ethanol with the required purity, but the bottoms product is a binary mixture of ethylene glycol and water. In this case it is important to highlight that an additional distillation column will be required to recover the ethylene glycol, and its energy consumption will be increased because the remixing effect is presented in the extractive distillation column. This inefficiency in the separation process is indicated in Figure 11; the composition of water increases until a maximum concentration at stage 18, but as the bottom of the distillation column is reached, the composition decreases drastically.

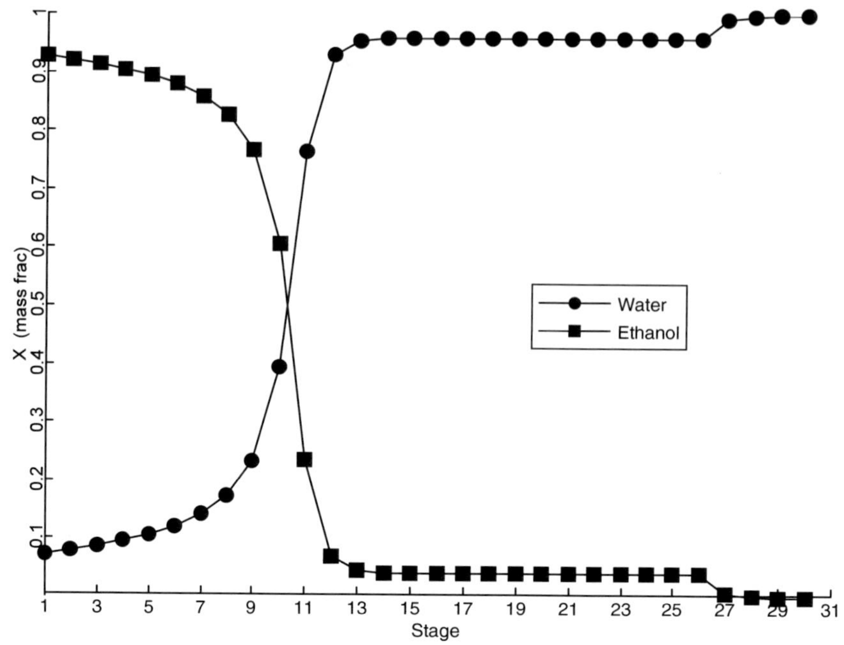

Figure 10. Composition profiles in the conventional distillation column.

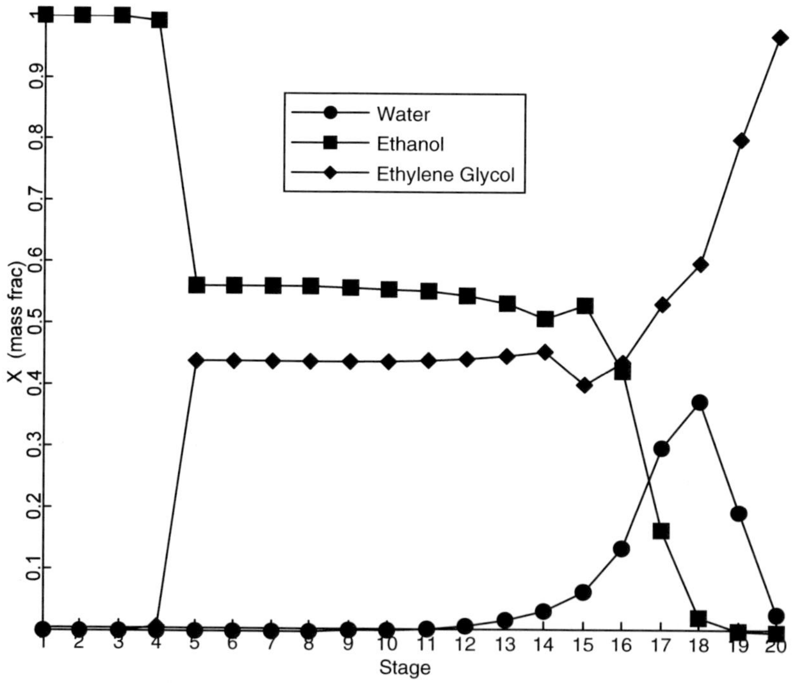

Figure 11. Composition profiles of the extractive distillation column.

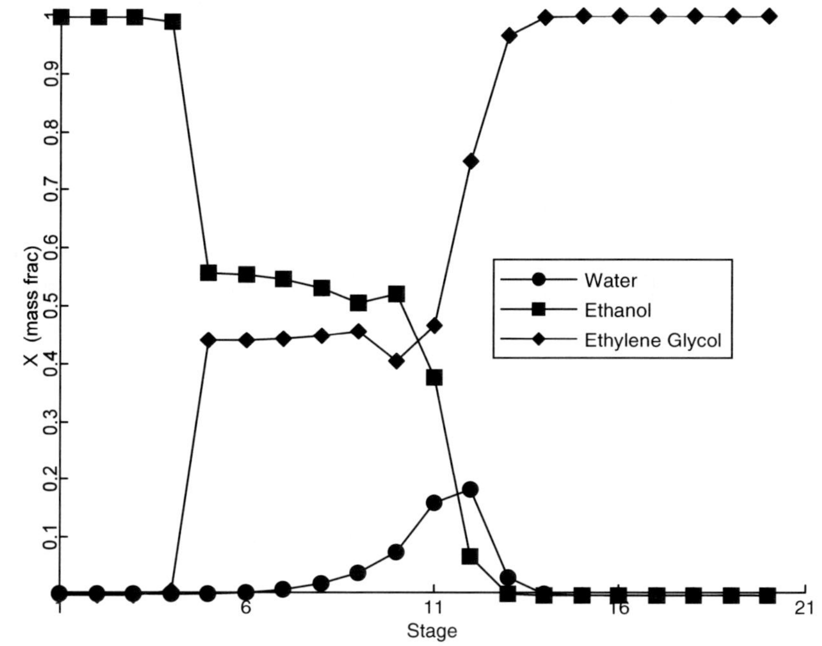

Figure 12. Composition profiles in the main column of the ETCDS-SR.

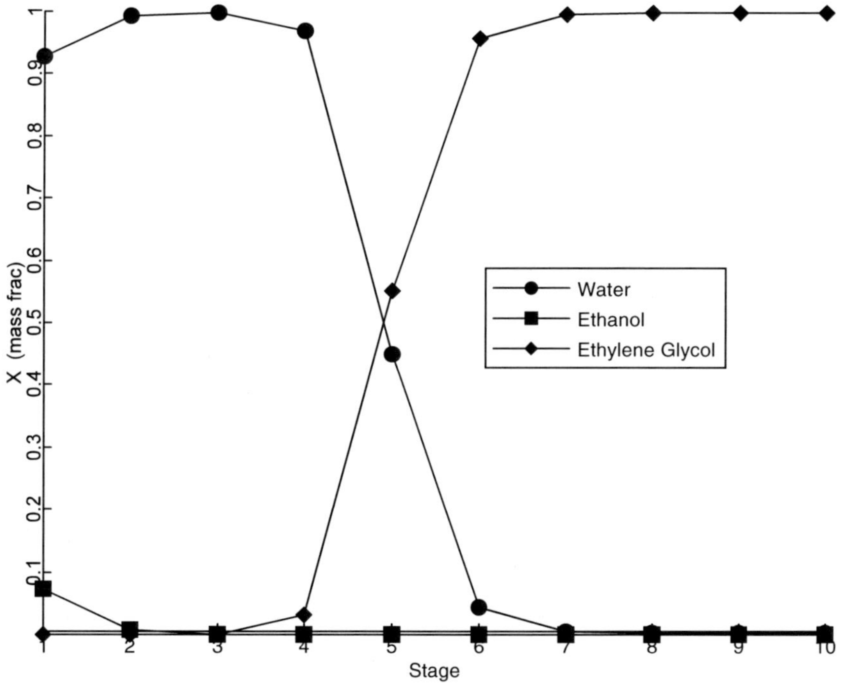

Figure 13. Composition profiles in the side rectifier of the ETCDS-SR.

In comparison with the separations achieved in the extractive distillation column of Figure 5, for the extractive complex distillation sequences of Figures 6-9, no additional distillation columns are required to recover ethylene glycol. To illustrate this, composition profiles of the extractive TCDS-SR are shown in figures 12 and 13 for the main and side rectifier columns, respectively. In the main distillation column (Figure 12), the top product is ethanol with the required purity, and the bottoms product is ethylene glycol. The side rectifier (Figure 13) has a dilute mixture of ethanol as top product that can be returned to the conventional distillation column (a distillation column that is common to the distillation sequences). The composition profiles for the distillation column with a sidestream and the distillation columns with one or two dividing walls are similar to those of the extractive ETCDS-SR, but the sidestream product is a binary mixture of ethanol and water that can be recycled to the first conventional distillation column. Again, the distillate is high purity ethanol and the bottoms product is the recovered ethylene glycol. According to the composition profiles of the distillation columns, where the ethylene glycol is used as entrainer, it can be concluded that this mass separation agent can be suitable for obtaining high purity ethanol since all the distillation sequences can adjust the composition to the desired value.

In order to test the other mass separation agents, the five distillation sequences were studied using a solution of NaCl with a mass concentration of 10%, but the composition of the ethanol in the product stream could not be achieved. For example, Figure 14 presents the composition profiles for the use of the solution of NaCl as entrainer; it can be noted that the composition of the product is around 0.98 mass fraction of ethanol. This value does not fit the purity constraint. As a result, it can be concluded that saline extraction cannot be used to obtain high purity bioethanol. Considering these results, the energy consumption and total annual costs were calculated only for the use of ethylene glycol as entrainer.

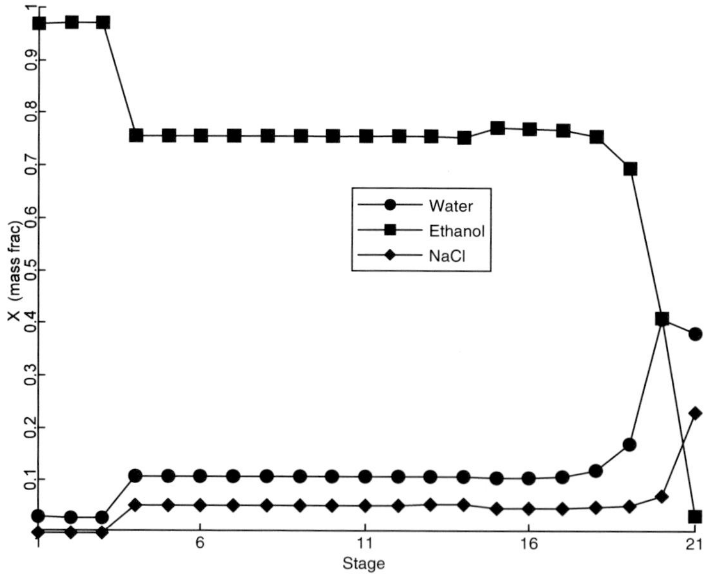

Figure 14. Composition profiles of the extractive distillation column.

The results regarding the energy consumption are shown in Table 2; it can be seen that the minimum energy consumption is presented by the conventional extractive distillation sequence. In addition, the extractive thermally coupled distillation sequence involving a side rectifier presents similar energy consumption. The other complex distillation sequences present energy consumptions that are 115% higher than those obtained in the best two options in terms of energy consumption (ECDS, ETCDS-SR). However, it is important to compare the extractive distillation sequences in terms of energy consumptions; this comparison does not take into account the temperatures of the utilities and the costs associated with the purchase, installation and operation of the extractive distillation sequences. In order to obtain a more realistic comparison, total annual costs (Equation 1) were obtained for the extractive distillation sequences (Table 2). According to the total annual costs, the best option is the extractive thermally coupled distillation sequence with a side rectifier, followed by the extractive distillation sequence with a sidestream.

$$\text{Total Annual Cost (TAC)} = \text{utility costs} + (\text{capital investment} / 5 \text{ years}) \quad (1)$$

Table 2. Total energy consumption (Q) and total annual cost (TAC) of the distillation sequences

Distillation sequence	Q (kW)	TAC (USD/year)
ECDS	314.8	291,553.41
ETCDS-SR	333.1	280,806.84
EDSS	524.2	387,228.90
EODWDC	607.9	390,373.64
ETDWDC	678.3	416,374.76

Finally, in accordance with the total annual costs, the best option for purification of bioethanol is the extractive thermally coupled distillation sequence with a side rectifier. This result is important because the total annual cost can be reduced further using a dividing wall for an industrial implementation. Figure 15 presents the proposed extractive distillation sequence using a dividing wall distillation column.

CONCLUSION

The process of obtaining high purity bioethanol from a dilute solution of a fermentation process was studied considering different extractive distillation sequences. The optimal designs of the extractive distillation sequences were obtained in a computational framework implemented in a commercial simulator and mathematical software using genetic algorithms. Ethylene glycol and a dilute solution of NaCl were considered to be mass separation agents in the extractive distillation sequences. Ethylene glycol was selected as a good option, since required purity of bioethanol could be achieved. Although NaCl is cheaper than ethylene glycol, this is not a suitable option because the composition of ethanol could not be adjusted to the desired value. The extractive distillation sequence with a side rectifier using ethylene glycol as entrainer was the best option in terms of total annual costs, and a practical implementation using a dividing wall is proposed.

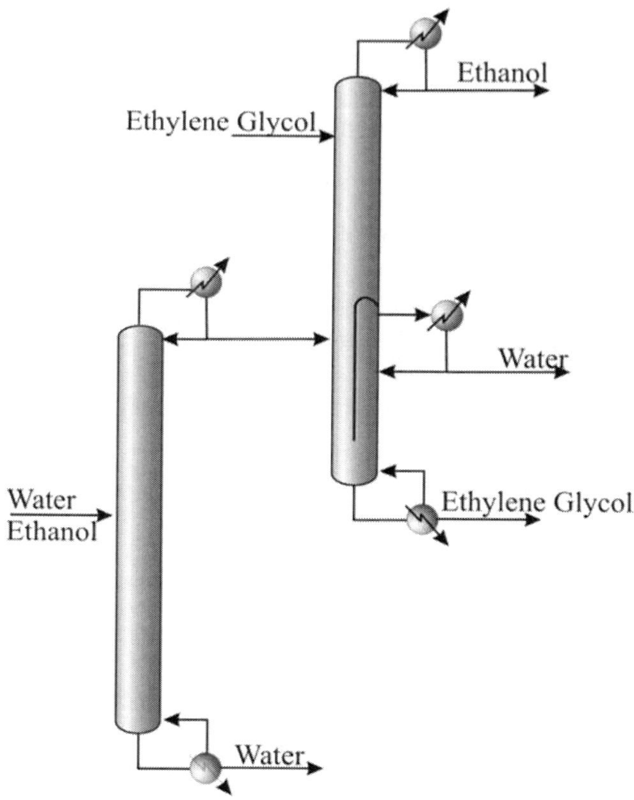

Figure 15. Practical implementation suggested for the ETCDS-SR.

Acknowledgements

We acknowledge the financial support provided by Universidad de Guanajuato, CONACyT and CONCyTEG (Mexico). The facilities given by the UMSNH during my sabbatical year is really appreciated.

References

[1] Quintero, J.A.; Montoya, M.I.; Sánchez, O.J.; Giraldo, O.H.; Cardona, C.A. *Energy,* 2008, 33, 385-399.

[2] Cardona, C. A.; Sanchez, O. J. *Bioresource Technology*, 2007, 98, 2415-2457.

[3] Wingren, A.; Galbe, M.; Zacchi, G. *Bioresource Technology*, 2008, 99, 2121-2131.

[4] Hernández, S.; Segovia-Hernández, J.G.; Rico-Ramírez, V. *Energy,* 2006, 31, 1840-1847.

[5] Mascia, M.; Ferrara, F.; Vacca, A.; Tola, G.; Errico, M. *Appl. Therm. Eng.,* 2007, 27, 1205-1211.

[6] Rong B.G.; Turunen, I. *Trans. Inst. Chem. Eng. Part A,* 2006a, 84, 1095-1116.

[7] Rong, B.G.; Turunen I. *Trans. Inst. Chem. Eng. Part A,* 2006b, 84, 1117-1133.

[8] Malinen, I.; Tanskanen, J. *Trans. Inst. Chem. Eng. Part A,* 2007, 85, 502-509.

[9] Tedder, D. W.; Rudd, D. F. *AIChE J,* 1978, 24, 303-315.

[10] Triantafyllou, C.; Smith, R. *Trans Inst. Chem. Eng.,* 1992, 70, 118-132.

[11] Kaibel, B.; Jansen, H.; Zich, E.; Olujic, Z. *In Distillation and Absorption '06*, Sorensen, E., *IChemE Symp. Series, IChemE, London,* 2006, 15, 252-257.

[12] Gutiérrez-Antonio, C.; Briones-Ramírez, A. *Computers and Chemical Eng*ineering, Accepted, 2008.

[13] Coello-Coello, C. A. *Civil Engineering and Environmental Systems*, 2000, 17, 319-346.

[14] Deb, K.; Agrawal, S.; Pratap, A.; Meyarivan, T. KanGAL report 200001, Indian Institute of Technology, 2000, Kanpur, India.

In: Bioethanol: Production, Benefits and Economics ISBN 978-1-60741-697-5
Editor: Jason B. Erbaum

Chapter 7

METHANOL CONTENT IN BIODIESEL ESTIMATED BY FLASH POINT AND ELECTRICAL PROPERTIES

S. D. Romano[1,2], P. A. Sorichetti[3] and I. Buesa Pueyo[1,4]

[1]Grupo de Energías Renovables, Facultad de Ingeniería,
Universidad de Buenos Aires, Av. Paseo Colón 850 (1063)
Buenos Aires, Argentina.
[2]CONICET (Consejo Nacional de Investigaciones Científicas y Técnicas),
Av. Rivadavia 1917 (1033) Buenos Aires, Argentina.
[3]Laboratorio de Sistemas Líquidos, Facultad de Ingeniería,
Universidad de Buenos Aires, Av. Paseo Colón 850 (1063)
Buenos Aires, Argentina.
[4]Universidad Politécnica de Valencia, Camino de Vera,
s/n 46022, Valencia, España.

ABSTRACT

International standards for Biodiesel (BD) characterization include two properties that are very important from the standpoint of safety in production, storage and transport of biofuels: Methanol Content (MC) and Flash Point (FP).

This chapter presents a systematic exploration of the direct relation existing between MC and FP, including electrical properties (permittivity and conductivity) that, although at present are not included in international standards for BD, provide relevant information on this subject.

Precisely known amounts of methanol (up to 2.5% V/V) were added to BD, and FP was determined by the well known Pensky-Martens method. An excellent fitting to a potential function was found between MC and FP. Moreover, the maximum allowable MC according to International Standards corresponds precisely to the minimum FP value indicated by the Standards.

Measurements of Electrical Properties (EP) carried out at different temperatures show a clear dependence between MC and the permittivity and conductivity of the samples. The results presented in this chapter show that FP and EP measurements are an interesting alternative for the verification of methanol content in biodiesel.

INTRODUCTION

Biodiesel (BD) is a liquid biofuel that can replace petroleum-based Diesel fuel (DF) or blended with it, since they can be mixed in any ratio. ASTM (formerly known as the American Society for Testing and Materials) defines biodiesel as a fuel comprised of mono-alkyl esters of long chain fatty acids derived from vegetable oils or animal fats, designated B100, and meeting the requirements of ASTM D 6751 [1].

In many countries, including the United States and the European Union, biodiesel in blends with DF in a proportion ranging from 5% (B5) to 20% (B20) is routinely used in Diesel engines without any modification. In consequence, world production and use of biodiesel has significantly increased in recent years, since it is a renewable fuel that offers an alternative to fossil fuels in automotive applications.

Biodiesel [2], [3] is obtained by transesterification of vegetable oils or animal fats with short linear - chain alcohol (methanol or ethanol). The reaction must take place under well defined conditions. Raw materials for BD production may include new or used, edible or non-edible vegetal oils and animal fats. Therefore, biodiesel may be produced from oil from soybean, sunflower, corn, peanut, cotton, palm tung, castor seed, jatropha, rapeseed, micro-algae, and animal fats from cows, pigs, chicken, fish, etc.

As a result of the transesterification process, glycerin and a mixture of esters are obtained. Biodiesel is obtained from this mixture of esters, after a purification process. It must be remarked that although the production process is straightforward, it is not always easy to obtain a final product that qualifies as BD according to the requirements of standards.

Quality control of fuel during production and distribution is essential to ensure the quality of the product delivered to users and the good performance of engines. To establish the good quality of the liquid biofuel, several properties must be measured. Moreover, since biodiesel properties depend on the type and proportion of esters, i.e. from the raw material, standards indicate allowable ranges for the key properties. To qualify as BD, measurement results must fall within these prescribed limits. Acceptance ranges for BD properties vary slightly between standards set by different organizations (ASTM D 6751, EN 14214, IRAM 6515, DIN V 51606, ÖN C 1191, UNI 10635, etc.)[1], [4], [5] but these differences have little technological impact. In some cases standards only establish maximum or minimum allowable values.

Parameters that define BD quality may be classified in two main groups [6]: general properties (including density, viscosity, flash point, cloud point, pour point, cetane number, acid number) and the chemical composition and purity of the mix of esters from fatty acids (alcohol content, ester content, proportion of mono-, di- and tri-glicerydes, total and free glycerin content, iodine index, etc.)

Although not all the standardization organizations require the measurement of the same properties, most of the required properties of BD are common to the different standards. The general properties of biodiesel used for characterization are: density, viscosity, lubricity, flash point, acid number, iodine index, cetane number, oxidation stability, carbon residue, sulfated ash, water content, non water soluble impurities and copper strip. The chemical composition and purity parameters include the content of methyl ester of linoleic acid, esters, methanol, monoglyceride, diglyceride and triglyceride, free glycerin, total glycerin, sulfur, alkaline metals and phosphorus.

The transesterification process may be basic, acid or enzymatic. At industrial scale, the basic process is employed, mainly using methanol. To achieve the highest yield, the amount of methanol exceeds the stoichiometric alcohol - oil ratio. Since it is an impurity, standards require that the free methanol content (MC) in BD be lower than 0.2 %. Therefore, the excess of alcohol must be removed from the mix of methyl esters by an adequate purification process. Flash point (FP) is an important parameter concerning the safety in BD distribution and storage, since higher values lead to a safer handling of the fuel. The minimum value required by standards is usually 100°C or higher.

It is very important to take into account that free methanol content and flash point are closely related. Since the FP of pure methanol is 11°C, it is not surprising that FP of biodiesel will be lower than 100°C if the MC exceeds 0.2%.

There are other properties, in addition to those required by the standards, that provide important information about the fuel, such as its heating value and electrical properties.

Specifically, measurement of electric properties is a widely used technique in liquid samples[7], [8]. Among other applications, electrical techniques make possible to characterize mixtures and detect contaminants independently from the color or turbidity of the liquid phases.

PREPARATION OF BIODIESEL SAMPLES

The samples of biodiesel studied in this chapter were prepared at the Grupo de Energías Renovables -GER- (Renewable Energy Group) at the Facultad de Ingeniería (Engineering School) of the University of Buenos Aires (UBA). Biofuel was made by the basic transesterification of new oil from soybeans and methanol, under constant stirring during one hour at a temperature of 60°C [7], [8]. Sodium hydroxide was used as a catalyst, and a 1:4 (in volume) alcohol to oil ratio. The mixture of methyl esters obtained as a product of the reaction was separated from glycerin and then purified by three washing steps. The first washing step was carried out with water acidified with hydrochloric acid and the remaining two with deionized water. Finally, samples were dried, in order to comply with the maximum water content value allowed by the standards [9].

In this chapter, the final product that complies with standards after the purification process is designated as "BD".

Characterization of Biodiesel According to International Standards

The samples of BD studied in this chapter were characterized by the following properties: viscosity, density, acid number, flash point (FP), cloud point, pour point, copper strip corrosion, free methanol content (MC) , free and total glycerin content and water content. All the measurement results were within the allowable ranges as required by international standards.

Preparation of Biodiesel Samples with Known Methanol Content

To obtain the value of FP as a function of MC, biodiesel samples prepared by the process described in the previous paragraphs were mixed with known volumes of methanol (analytical grade).

DETERMINATION OF FLASH POINT

Flash point measurements were carried out using the Pensky-Martens Closed Cup Tester according to ASTM D 93 [10]. An outline of the Pensky-Martens apparatus, in accordance with the design illustrated in ASTM D93, is shown in Figure 1.

The test sample is placed in a brass test cup of specified dimensions that is fitted with a cover of specified dimensions. The cup is heated and the sample is stirred at a specified rate. At regular intervals, the stirring is interrupted and at the same time an ignition source is directed into the test cup, until a flash is detected. The flash point is defined as the lowest temperature at which application of an ignition source causes the vapors of the test sample to ignite under the specified conditions indicated by the standard. For the purposes of FP determination, it is considered that the test sample has flashed when a flame appears and instantaneously propagates itself over the entire surface.

The following test conditions are set by the Standard:

- The temperature of the sample must increase at a rate of 5 to 6°C/minute.
- The ignition source must be applied at each temperature increase of 2°C, beginning at a temperature of 23 +/- 5°C below the expected flash point.
- The required sample volume is 70 ml.

It should be noted that, according to most standards, FP of biodiesel should be higher than 100°C, and usually it is above 130°C. This is a significant difference with the FP of Diesel fuel, which is above 50°C, and it is one of the advantages for the transport and storage of BD.

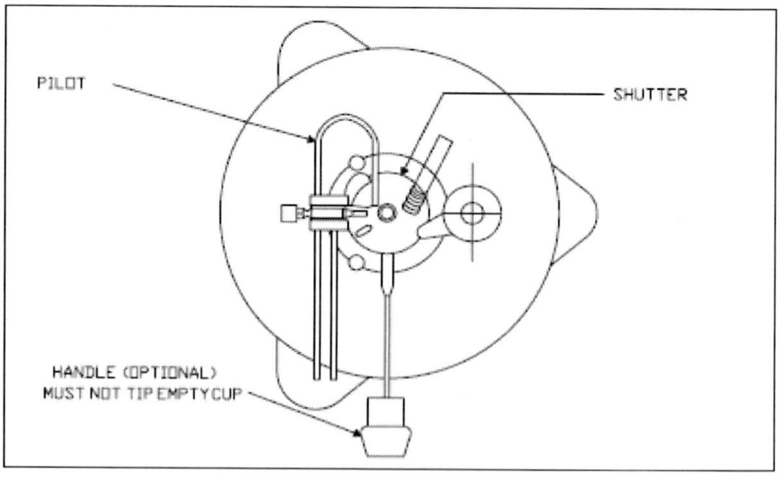

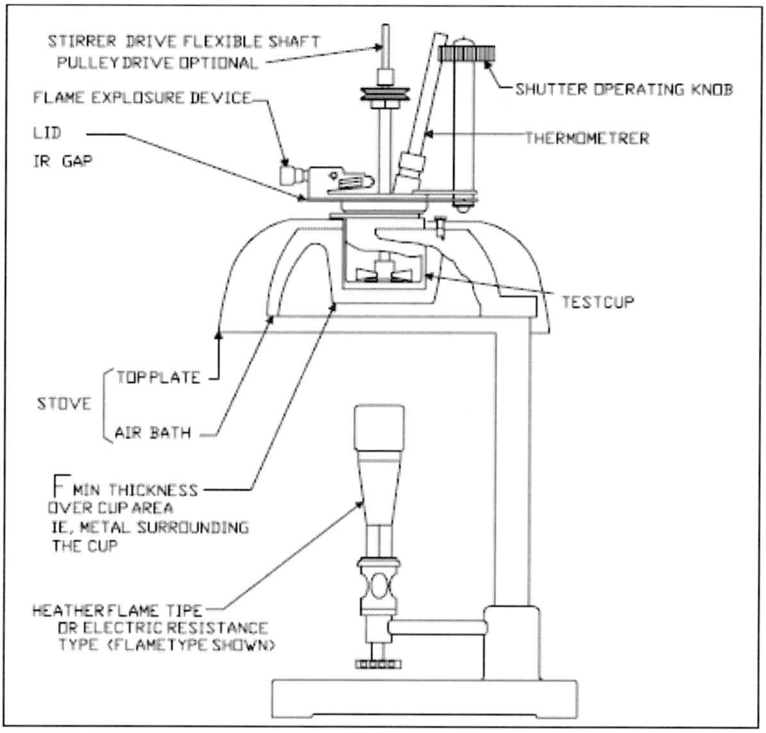

Figure 1. Scheme of Pensky - Martens Closed Flash Tester Equipment.

MEASUREMENT OF PERMITTIVITY AND CONDUCTIVITY IN LIQUIDS

Broadly speaking, the relative permittivity of a substance characterizes its dielectric polarization, that is, the macroscopic (average) response of the permanent and induced dipole moments at the atomic and molecular scales, when an external electric field is applied. Similarly, conductivity determines the macroscopic current originated by transport processes of free charges (electrons and ions), when the substance is subjected to an external electric field [11].

The conductivity and permittivity of the sample as a function of the excitation angular frequency, ω, are usually jointly described by the complex relative permittivity $\varepsilon_r(\omega)$, defined as the ratio of the capacitance of the cell containing the sample, $C_s(\omega)$, to the capacitance of the empty cell, C_0.

$$\varepsilon_r(\omega) = \frac{C_s(\omega)}{C_0} \quad (1)$$

where the imaginary part of $\varepsilon_r(\omega)$ is written with a negative sign:

$$\varepsilon_r(\omega) = \varepsilon_r'(\omega) - i\varepsilon_r''(\omega) \quad (2)$$

As follows from (1) the real part of $\varepsilon_r(\omega)$ is proportional to the electrostatic energy stored as dielectric polarization, whereas the imaginary part $\varepsilon_r''(\omega)$ represents the energy lost through dissipative processes. Usually, the imaginary part of the permittivity is often written as the sum of two terms:

$$\varepsilon_r''(\omega) = \varepsilon_{r\,pol}''(\omega) + \frac{\sigma}{\varepsilon_0 \omega} \tag{3}$$

The first term, $\varepsilon_{r\,pol}''(\omega)$ represents the dissipative processes (relaxation) associated to the dielectric polarization. Therefore, it is related to the same molecular and electronic processes that give rise to the dipole moments described by the real part of the permittivity, $\varepsilon_r'(\omega)$. The second term describes the energy loss associated to the electrical conduction due to free charge carriers (ions and electrons).

It is proportional to the conductivity parameter, σ, and inversely proportional to the angular frequency of the excitation, ω. Clearly, at low frequencies ($\omega \to 0$) conductivity often is the most important dissipative process and therefore dominates the imaginary part of complex permittivity. The constant ε_0 is the permittivity of free space (8.854 10^{-12} F/m).

The phase angle (with a minus sign) of the complex permittivity as a function of ω is customarily designated as $\delta(\omega)$. It is also usual in technological applications to define the dissipation factor $D_f(\omega)$ as the tangent of $\delta(\omega)$, that is, as the ratio (with a minus sign) of the imaginary to the real part of the complex permittivity $\varepsilon_r(\omega)$:

$$D_f(\omega) = tg\,\delta(\omega) = \frac{\varepsilon''_r(\omega)}{\varepsilon'_r(\omega)} \tag{4}$$

It is easy to see from equations. (1) and (2) that, under steady - state time - harmonic excitation, the dissipation factor is proportional to the ratio of the energy dissipated to the energy stored per cycle in the dielectric.

The results of complex permittivity measurements are usually presented as plots of the real and imaginary parts, $\varepsilon_r'(\omega)$ and $\varepsilon_r''(\omega)$. The dissipation factor $D_f(\omega)$ is often given, since in the frequency range where free charge transport is the predominant dissipative process, $D_f(\omega) \sim \omega^{-1}$, which is easily recognized in a log-log graph of $D_f(\omega)$.

Measurements of permittivity and conductivity are based in the interaction of the substance under study with a time-varying electric field. The liquid sample is introduced in a cell kept at constant temperature, usually by means of a thermostatic system. Inside the cell, a set of electrodes connected to the measuring circuit are submerged in the liquid. In the measuring range of interest in this chapter (20 Hz to 2 MHz), the measuring circuit provides a constant-amplitude, time-harmonic excitation to the cell. The response signal is processed and compared the excitation signal, in order to determine the capacitance of the cell, as a function of frequency. Measurement of capacitance may be carried out using stand-alone instruments made by different manufacturers, such as RCL Meters, Impedance Analyzers and Dielectric Analyzers [12], [13], [14].

As an alternative, other dielectric measuring systems use separate instruments: a Signal Generator to provide the excitation, a Dielectric Measuring Interface connected to the sample cell, and to compare the response signal with the excitation, a Phase-Amplitude Analyzer, a Lock-in Amplifier or a Digital Oscilloscope controlled by a Personal Computer (PC) with Fourier Analysis (FFT) software.

In the authors experience, sample cells with platinized platinum electrodes (of the usual type used for electrochemical work), are quite adequate for the measurements described in this chapter.

Measurements presented in this chapter were carried out by an automated dielectric system [15], [16] consisting of a GW - 830 Signal Generator, a custom - design Dielectric Measuring Interface and a Tektronix TDS - 210 Digital Oscilloscope. The temperature of the sample cell was controlled by a Lauda thermostat.

DETERMINATION OF THE TEMPERATURE DEPENDENCE OF ELECTRICAL PROPERTIES OF BIODIESEL

In the frequency range of interest in this chapter, from about 20 Hz to 2 MHz, no dielectric relaxation processes are observed in biodiesel; therefore, ε_r' may be considered as independent of ω and $\varepsilon_{r\ pol}''$ is negligible. On the other hand, the presence of free charge carriers due to the presence of contaminants and thermal effects in BD originate dissipative effects and the conductivity term σ/ω must be included in the imaginary part of the complex permittivity. Moreover, since thermal agitation tends to oppose the polarization associated to molecular orientation, ε_r', may be expected to decrease with temperature.

In summary, in this chapter the dependence of electrical properties of BD with the absolute temperature T and the angular frequency of the excitation, ω, will be fitted to the following expression:

$$\varepsilon_r(\omega,T) = \varepsilon_r'(T) - \mathrm{i}\,\frac{\sigma(T)}{\varepsilon_0\ \omega} \tag{5}$$

The real part $\varepsilon_r'(T)$ will be assumed to have a linear dependence on temperature

$$\varepsilon_r'(T) = a - b\ T \tag{6}$$

where the parameters a and b depend on the BD composition and also are affected by the presence of contaminants.

On the contrary, the conductivity of pure BD is very low but increases exponentially with temperature [8]; moreover, the presence of contaminants changes significantly the room temperature value of conductivity and it dependence with the temperature.

CORRELATION BETWEEN FLASH POINT AND METHANOL CONTENT IN BIODIESEL

Measurements show that there is a strong correlation between flash point and methanol content (Figure 2). The following expression fits very well the experimental data, with a correlation coefficient (R^2) greater than 0.99:

$$y = 38 \quad x^{-0.6} \tag{7}$$

where x is the methanol content in biodiesel (% V/V) and y is the flash point of the sample (°C), with $x \geq 0.2$. FP of BD samples with no methanol added always exceeded the maximum FP value that could be measured with the equipment available at the GER (> 180°C).

All the FP measurements were double-checked.

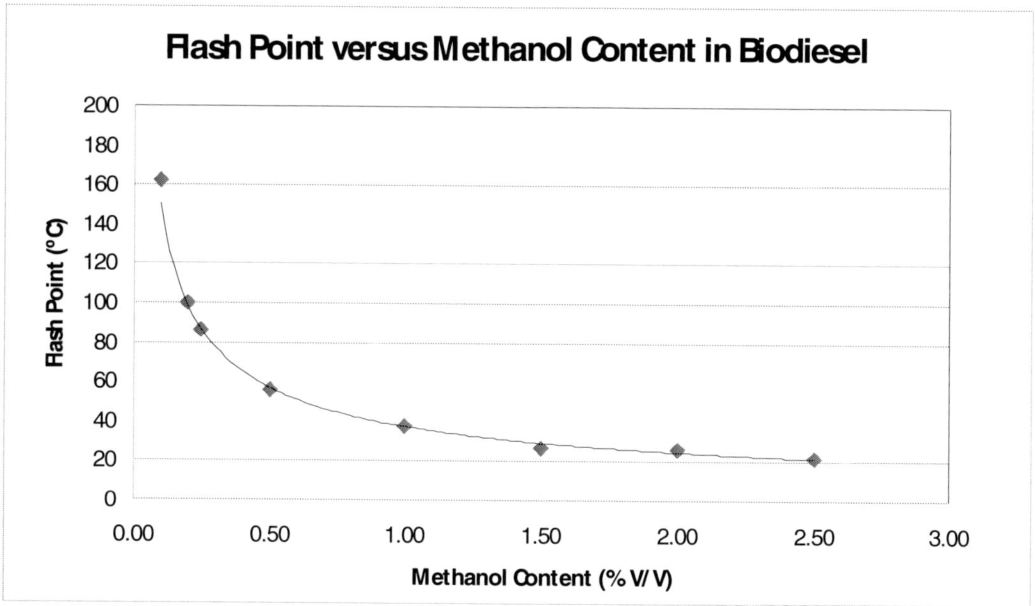

Figure 2. Flash Point versus Methanol Content in Biodiesel.

It is important to remark that the measured flash point value at the maximum methanol content set by international standards (0.2%) corresponds to the minimum allowable flash point (100°C) according to the standards. Therefore, compliance with the flash point requirement is also a good indicator of compliance with maximum methanol content. As it is easy to see in Figure 2, a small excess in methanol content originates a sharp decline in flash point of biodiesel.

PERMITTIVITY AND CONDUCTIVITY IN BIODIESEL CONTAINING METHANOL

Regarding electrical properties, experimental data show that biodiesel containing methanol has a higher permittivity at room temperature than pure BD [7], [8]. As indicated in equation (6), the dependence of the real part of the complex permittivity, $\varepsilon_r'(T)$, may be fitted to a linear function, where the parameters *a* and *b* depend on methanol concentration.

From the measurement results plotted in Figure 3, the fitting parameters are summarized in Table I.

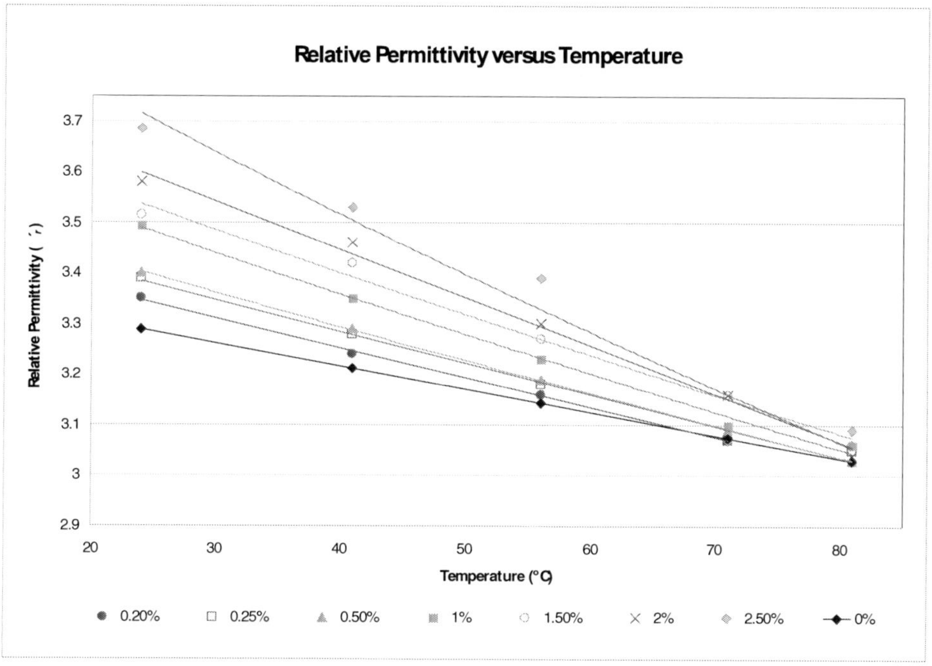

Figure 3. Relativity Permittivity versus Temperature as a function of methanol content in Biodiesel.

Table I. Parameters of the linear fit and correlation coefficient for permittivity versus temperature at different values of methanol content in biodiesel

Methanol concentration in Biodiesel (%V/V)	a	b	R^2
0.00	3.398	0,0046	0,999
0.20	3.490	0,0059	0,998
0.25	3.534	0,0062	0,988
0.50	3.544	0,0061	0,998
1.00	3.645	0,0074	0,994
1.50	3.747	0,0084	0,996
2.00	3.828	0,0095	0,993
2.50	3.997	0,0117	0,949

From Figure 3 it may be seen that the rate of decrease of permittivity with temperature increases with methanol content. Also, the fitting of permittivity data as a linear function of temperature is better at lower methanol concentrations (c.f. Table I). At higher temperatures $\varepsilon_r'(T)$ tends to the value of pure BD. This is not surprising since methanol evaporates at 65°C.

From figure 4, below 65°C the conductivity of biodiesel increases with higher methanol content.

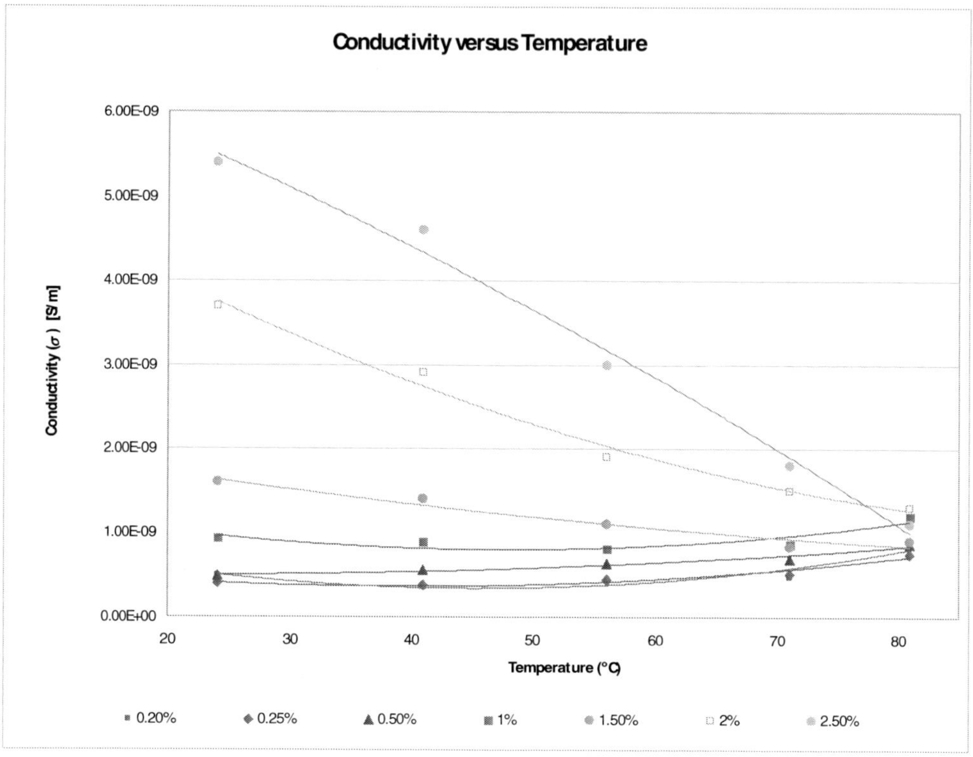

Figure 4. Conductivity versus Temperature as a function of methanol content in Biodiesel.

CONCLUSION

Measurement data of flash point as a function of methanol content show clearly that small increases of alcohol concentration originate a steep decline in FP. This result is important for production, distribution and utilization of BD. In particular, excess of methanol must be eliminated with an adequate purification process.

From the strong correlation between MC and FP, it is possible to use FP measurements for the preliminary verification (prior to certification) of the compliance with the methanol content requirements of BD standards. This is relevant since FP measurements are much simpler and use less expensive equipment than the procedure for determination of methanol content recommended by standards (chromatographic analysis) [17].

The linear fit of permittivity measurements as a function of temperature makes possible to determine accurately MC in biodiesel within a wide range of concentrations, including the maximum value set by the standards (0.2%). Methanol content may be obtained from the value of $\varepsilon_r'(T)$ extrapolated to 0°C, that is, the parameter *a* in equation (6), given in Table I.

Measurements of biodiesel conductivity at room temperature (25°C) may be used for a preliminary check of methanol concentrations above 0.5%, as shown in Figure 4. This is very convenient since the equipment for conductivity measurement at room temperature is widely available and easier to use than systems for permittivity measurements as function of temperature, particularly in industrial applications.

Acknowledgements

One of the authors (Buesa Pueyo) thanks Universidad Politécnica de Valencia (Spain) for a PROMOE scholarship to carry out his Graduation Project in Agricultural Engineering at the Renewable Energy Group (GER) of the University of Buenos Aires.

The work in this chapter was funded by Project UBACYT I409 of the University of Buenos Aires: "Optimization of process variables for biodiesel production in an automated pilot plant, from different alternative raw materials".

References

[1] ASTM D 6751- 03 a (American Standard for Biodiesel): Standard Specification for Biodiesel Fuel Blend Stock (B100) for Middle Distillate Fuels, 2003.

[2] Knothe G, Van Gerpen J, and Krahl J. *The Biodiesel Handbook*. AOCS Press, 2005.

[3] Romano, S. D., González Suárez, E., and Laborde, M. A. *Combustibles Alternativos*. Ediciones Cooperativas, 2nd Ed. 2006. Chapter I: Biodiesel, S. D. Romano, pp. 11 – 88.

[4] EN 14214 (European Standard for Biodiesel): Automotive Fuels, Fatty Acid Methyl Esters (FAME) for Diesel Engines, Requirements and Test Methods.

[5] IRAM 6515-1/02 (Argentine Standard for Biodiesel): Calidad de Combustibles. Combustibles líquidos para uso automotor. *Biodiesel,* 2002.

[6] Meher, L. C.; Vidya Sagar, D.; Naik, S. N. *Renewable and Sustainable Energy Reviews*, 2006, *10*, 248 – 268.

[7] Sorichetti, P. A.; Romano, S. D. *Physics and Chemistry of Liquids*, 2005, *43 (1)*, 37 – 48.

[8] González Prieto, L. E.; Sorichetti, P. A.; Romano, S. D. *International Journal of Hydrogen Energy*, 2008, *33*, 3531 – 3537.

[9] ASTM D 4928 – 00: Standard Test Methods for Water in Crude Oils by Coulometric Karl Fischer Titration, 2000.

[10] ASTM D 93 – 02 a: Standard Test Methods for Flash Point by Pensky-Martens Closed Cup Tester, 2002.

[11] von Hippel, A. R.. *Dielectric and Waves*, John Wiley and Sons, New York, 1st Edition, 1954, Chapter 1.

[12] Field, R.F. *Dielectric Materials and Application*; von Hippel, A.R; The MIT Press, Cambridge, Mass., 1966, Ch. 2, 47 – 62.

[13] van Roggen, A. IEEE Trans. on Elec. Insulation, 1990, *25*, 95-106.

[14] Kremer, F. ; Arndt, M., *Dielectric Spectroscopy of Polymeric Materials*; Runt, J. P. and Fitzgerald, J. J., American Chemical Society, Washington D.C., 1997, Ch. 2, 67 - 74.
[15] Schenkel, C. D.; Sorichetti, P. A.; Romano, S. D. Anales de la Asociación Física Argentina, 2005, *17*, 283 - 287.
[16] Sorichetti, P. A.; Matteo, C. L. *Measurement*, 2007, *40 (4)*, 437 - 449.
[17] EN-14110: Fat and oil derivatives – Fatty Acid Methyl Esters (FAME) – Determination of methanol content.

In: Bioethanol: Production, Benefits and Economics
Editor: Jason B. Erbaum
ISBN 978-1-60741-697-5

Chapter 8

CORN ETHANOL ALLURE PLUMMETS

David Pimentel
Insect Ecology and Agricultural Sciences,
Department of Entomology andSection of Ecology and Systematics,
Cornell University

The diminishing supply of oil and high prices is encouraging the conversion of grain or other biomass into ethanol fuel. Using corn or any other biomass for ethanol requires large land areas of fertile soil, and sunlight for green plant production plus significant quantities of water. All the green plants in the U.S., including all crops, forest, and grasses, combined collect only about 32 quads (32 x 10^{15} BTU) of sunlight energy per year. Meanwhile, the American population uses slightly more than 3 times that amount of energy each year as fossil fuels [USCB, 2007]!

Ethanol supporters claim that ethanol could replace much of the oil used in U.S. However, consider that in 2006, 20% of the U.S. corn crop was converted into 6 billion gal of ethanol, but this amount replaced only 1% of U.S. oil consumption [USCB, 2007]. If the entire corn crop were used, it would replace a mere 7% of oil consumption–and not make the U.S. independent of foreign oil!

Up-to-date analyses confirm that 14 energy inputs typically are required for corn production, then 9 more energy inputs are invested in fermentation and distillation operations, confirming that more than 146% more energy (mostly high value oil and natural gas) is expended to produce 1 gallon of corn ethanol than is in the ethanol gallon itself [Pimentel and Patzek, 2008]. Some investigators omit several of the energy inputs required in corn production and processing, such as energy for farm labor, farm machinery, energy used producing hybrid corn-seed, irrigation, and processing equipment [Shapouri et al., 2002; Shapouri et al, 2004]. Omitting several fossil energy inputs suggests that a corn ethanol production system provides a positive energy return. Clearly corn ethanol is an inefficient choice from an energy cost and transport standpoint.

The production of corn ethanol is highly subsidized by state and federal governments by more than $6 billion per year according to a 2006 report, "Biofuels --at What Cost? Government Support for Ethanol and Biodiesel in the United States," released by the International Institute for Sustainable Development in Geneva [Koplow, 2006]. These current subsidies for a gallon of ethanol are more than 60 times those for a gallon of gasoline.

The environmental impacts of corn ethanol are numerous and serious. These include severe soil erosion of valuable cropland, plus the heavy use of nitrogen fertilizer and pesticides that pollute rivers [NAS, 2003]. Large quantities of carbon dioxide are produced and released into the atmosphere as significant amounts of fossil fuel energy are used in ethanol production [Pimentel and Patzek, 2008]. Then, during the fermentation process, about 25% of the carbon from the sugars and starches is released as carbon dioxide into the atmosphere. These two major releases of carbon dioxide significantly contribute to global warming.

Each gallon of corn ethanol requires 1,700 gallons of water (mostly to grow the corn) and releases 12 gallons of noxious sewage effluent from the fermentation process into the environment [Pimentel and Patzek, 2008].

Cellulosic ethanol is touted as the replacement for corn ethanol. Although 22 pounds of corn grain is required to produce 1 gallon of ethanol, depending on the cellulosic biomass, anywhere from 66 to 110 pounds of cellulosic biomass is required to provide as much starches and sugars as 22 pounds of corn [Pimentel and Patzek, 2008]. Another problem with cellulosic biomass is that the starches and sugars are tightly bound in the biomass and major fossil energy inputs are needed to release the tightly bound starches and sugars for ethanol conversion. About 170% more energy (oil and gas) is required to produce ethanol from cellulosic biomass than the ethanol produced [Pimentel and Patzek, 2008].

Biodiesel produced from soybeans is another popular biofuel from another plant group. Soybeans are the best oil producing crop for biodiesel production because soybeans can be produced without the application of nitrogen fertilizer. Applied nitrogen makes up about one-third the total energy required in corn production [Pimentel and Patzek, 2008], thus omitting the nitrogen in soybean production is a major benefit. However, the problem with soybean production is the relatively low yield compared with corn grain [USDA, 2006; USCB, 2004-2005]. Corn grain yields about 9,000 pounds per acre whereas soybeans yield only about 2,500 pounds per acre. Thus, extracting the 18% oil contained in soybeans provides a relatively small amount of oil per acre. The fossil energy input for soy, compared with the biodiesel produced, results in a negative energy return of about 60% as fuel [Pimentel and Patzek, 2008].

Some workers have suggested that algae cultures can supply the U.S. with all of its oil needs. One study suggests based on a paper study that up to 33,000 gallons/acre of oil could be produced using algae [Pimentel, 2008]. If the above estimated production and price of oil produced from algae were exact, U.S. annual oil needs could theoretically be met if 100% of all U.S. land were in algal culture [Pimentel, 2008].

Despite all the algae-related research and claims dating back to 1970's, none of the projected algae and oil yields has been achieved. To the contrary, one calculated estimate based on all the included costs using algae would be $20 per gallon, not $1 per gallon [Pimentel, 2008]. Algae, like all plants, require large quantities of nitrogen and water, in addition to the significant fossil energy inputs required for the production system. Thus, a great deal of research is needed before the proposed use of algae for oil production is confirmed.

Using food crops, such as corn grain, to produce ethanol also raises major nutritional and ethical concerns. Nearly 60% of humans in the world now are currently malnourished [WHO, 2005], so the need for grains and other basic foods is critical. Growing crops for fuel squanders land, water, and energy resources vital for the production of food for people. The

President of the World Bank reported that biofuels have increased world food prices 75% [Chakrabortty, 2008]. Jacques Diouf, Director General of the U.N. Food and Agriculture Organization reports biofuels are increasing world human starvation [Diouf, 2007].

The world population is currently at 6.7 billion with a quarter million additional people added daily [PRB, 2007]. Energy specialists project that peak oil has already been reached and there are only about 60 years of this fuel remaining [BP, 2005]. There should be about 100 years of natural gas remaining. Slowly oil and gas supplies will decline until these fuels are exhausted. This will create a critical situation for food production because all food supply currently depends primarily on oil and gas to maintain a highly productive agriculture. The Food and Agriculture Organization reports that cereal grain production per capita has been declining continuously for the past 24 years [FAO, 1961-2008]. This is critical because grains make up 80% of world food. Food and biofuels are dependent on the same resources for production: land, water, and energy. Increased use of biofuels further damages the global environment and especially the world food system.

REFERENCES

BP. (2005).*BP Statistical Review of World Energy.* London: British Petroleum.

Chakrabortty, A. (2008). Secret report: biofuel caused food crisis: Internal World Bank study delivers blow to plant energy drive. *The Guardian.* Friday July 4 2008. http://www.guardian.co.uk/environment/2008/jul/03/biofuels.renewableenergy (11/24/08).

Diouf, J. (2007). Biofuels a Disaster for World Food. In EU Coherence: EU Coherence Policy for Development. http://eucoherence.org/renderer.do/clearState/false/menuld/227351/return (10/31/07).

FAO. (1961-2006). Food and Agricultural Organization of the United Nations. Rome: FAO.

Koplow, D. (2006). Biofuels – at what cost? Government support for ethanol and biodiesel in the United States. The Global Studies Initiative (GSI) of the International Institute for Sustainable development (IISD). http://www.globalsubsidies.org/IMG/pdf/ biofuels_subsidies_us.pdf (2/16/07).

NAS. (2003). Frontiers in agricultural research: Food, health, environment, and communities. Washington, DC: National Academy of Sciences. http://dels.nas.edu/rpt_briefs/ frontiers_in_ag_final%20for%20print.pdf (11/05/04)

Pimentel, D. (2008). A brief discussion on algae for oil production: Energy issues. In D. Pimentel (Ed.), *Biofuels, solar and wind as renewable energy systems: Benefits and risks* (pp. 499-500). Dordrecht, The Netherlands: Springer.

Pimentel, D. and Patzek, T. (2008). Ethanol production using corn, switchgrass and wood; Biodiesel production using soybean. In D. Pimentel (Ed.), *Biofuels, solar and wind as renewable energy systems: Benefits and risks* (pp.373-394) Dordrecht, The Netherlands: Springer.

Pimentel, D., Marklein, A., Toth, M.A., Karpoff, M., Paul, G.S., McCormack, R., Kyriazis, J., and Krueger, T. (2008). Biofuel impacts on world food supply: Use of fossil fuel, land

and water resources. *Energies* 1, 41-78; DOI: 10.3390/en1010041, online, open access. http://www.mdpi.org/energies/papers/en1020041.pdf

PRB. (2007). World Population Data Sheet, 2007. Washington, DC: Population Reference Bureau.

Shapouri, H., Duffield, J.A., and Wang, M. (2002). The energy balance of corn ethanol: An update. Report 813. Washington, DC: USDA, Office of Energy Policy and New Uses, Agricultural Economics.

Shapouri, H., Duffield, J., McAloon, A., and Wang, M. (2004). The 2001 net energy balance of corn-ethanol (Preliminary). Washington, DC: U.S. Department of Agriculture.

USCB. (2007). *Statistical Abstract of the United States, 2007.* U.S. Census Bureau. Washington, DC: U.S. Government Printing Office.

USDA. (2006). *Agricultural Statistics, 2006.*. U.S.Department of Agriculture. Washington, DC: U.S. Government Printing Office.

WHO. (2005). Malnutrition worldwide. World Health Organization. http://www.mikeschoice.com/reports/malnutrition_worldwide.htm. (12/7/2007).

In: Bioethanol: Production, Benefits and Economics
Editor: Jason B. Erbaum
ISBN 978-1-60741-697-5

Chapter 9

BIOFUELS

The worldwide depletion of fossil fuels and widespread concern over increasing atmospheric CO2 have sparked interest not only in biomaterials but also in sustainable, non-fossil-based fuels. Political instability in petroleum-producing regions has further increased the desirability of domestic fuel sources, particularly for transportation [1]. Solar and wind power are well-suited to sustainable generation of electricity, including electricity for charging vehicle batteries, but most modern vehicles are designed for liquid fuels that are best simulated by two biofuels: bioethanol and biodiesel [2]. In addition, biohydrogen is an emerging biofuel that carries energy from sunlight or organic matter, rather than petroleum, in clean-burning hydrogen (H_2), Finally, biodesulfurization of petroleum products may offer a way to mitigate some effects of petroleum use during a transition and is discussed as well.

A. BIOETHANOL

1. Introduction

Proponents of ethanol as a use for fuel highlight the apparent net-zero contribution of fuel ethanol combustion to the global carbon cycle, in that feedstocks for ethanol production derive their carbon from atmospheric CO_2, and that ethanol combustion simply returns the fixed carbon to its atmospheric source [3]. Others, however, make the valid counterpoints that the conversion even of agricultural wastes to ethanol is, itself, an energy-intensive process that frequently makes use of fossil energy sources, and that growth of crops dedicated to energy production must also be conducted in a sustainable manner for fuel ethanol use to carry a net environmental benefit [4].

Petroleum currently supplies 97 percent of the energy consumed for transportation [5], and transportation accounted for two-thirds of U.S. petroleum use in 2002. This trend is expected to continue until 2025 [6]. This need not continue, however, as all automobile manufacturers produce flexible-fuel vehicles (FFVs) that can use 10 percent or 85 percent ethanol blends with gasoline, and ethanol can also replace diesel fuel in heavy vehicles [5]. The United States also now has 199 fueling stations for ethanol, as well as extensive online services for planning travel between stations [7]. Although ethanol is limited in availability in some states, the transportation market for ethanol could expand to as much as 3 8–53 billion liters per year, if all available agricultural residues were converted to ethanol [8]. Ethanol is also being used as a replacement for methyl tertiary butyl ether (MTBE), the fuel oxygenate that is being phased out due to its widespread contamination of groundwater [9].

The majority of ethanol, approximately 62 percent of the world total, is currently produced in Brazil, primarily from cane sugar (~12.5 billion liters in 2002), and in the United States, primarily from corn (~5 billion liters per year) [5, 10]. However, these feedstocks are expensive and are useful as foods, causing a great deal of research to be focused on the development of biomass such as corn cobs and stalks, sugar cane waste, wheat and rice straw, other agronomic residues, forestry and paper mill discards, paper municipal waste, and dedicated energy crops into ethanol [11]. While the use of such non-food substrates helps the economics of ethanol production substantially, the high cost of production, especially relative to gasoline, remains the primary obstacle to bioethanol commercialization [5].

Lignocellulosic (non-food) raw materials, such as agricultural, wood chip, and paper wastes, can yield approximately 100 billion gallons of fuel-grade ethanol per year in the United States alone [12]. The projected cost of bioethanol has dropped from about $1.22 per liter to about $0.31 per liter based on consistent improvements in pretreatment, enzyme application, and fermentation [13]. If additional specific improvement targets are met, this cost could drop to as low as $0.20–$0.12 per liter by 2015 [14]. For transportation fuel, therefore, ethanol has real potential to replace gasoline, even in the absence of governmental support.

2. State of the Science

2.1. Ethanol Biosynthesis: Overview

2.1.1. Feedstocks. Biological production of ethanol first requires that atmospheric CO_2 be fixed into organic carbon (biomass) through photosynthesis. While agricultural crops and residues currently form the vast majority of feedstock for ethanol production, other biomass sources such as wood chips, sawdust, industrial organic wastes, and municipal organic wastes are important for the commercial development of fuel bioethanol [11]. Agricultural plant matter contains approximately 10–15 percent lignin, a polymer of phenolic subunits that is highly resistant (although far from impervious) to enzymatic attack. Lignin typically surrounds and protects the more enzymatically-vulnerable components of cellulose and hemicellulose, which comprise approx. 20–50 percent and 20–3 5 percent of the remaining plant material, respectively [9].

The next challenge in bioethanol synthesis is therefore the release of fermentable sugars from the biomass, or conversion of the feedstock into fermentable substrates. This phase involves both the separation of lignin from cellulosic and hemicellulosic polymers and the hydrolysis of the polymers into monomeric sugars, primarily glucose and xylose. This phase is termed pretreatment, and a variety of biotic and abiotic approaches are currently under investigation; abiotic approaches have been recently reviewed [15].

2.1.2. Mechanical and chemical disruption. Lignin is typically dissociated from the carbohydrates by mechanical and/or thermochemical means, including hot water, steam explosion, and/or acid treatments in either batch or flow-through reactors. Many variations have been explored. While it was once difficult to compare the performance and economics of the various approaches due to differences in feedstocks tested, a group of pretreatment researchers has formed in North America to facilitate such comparisons. This group, the

Biomass Refining Consortium for Applied Fundamentals and Innovation (CAFI), has the goal of advancing the efficacy and knowledge base of pretreatment technologies (3), reviewed in [16, 17].

Mechanical lignin disruption effectively hydrolyzes a significant fraction of the hemicellulose, but is less effective in hydrolyzing cellulose. This difference is caused by the different structures of the two polymers: hemicellulose is a highly branched, typically amorphous polymer that is therefore relatively easy to hydrolyze into its component sugars (pentoses D-xylose and L-arabinose; hexoses D-galactose, D-glucose, D-mannose; and uronic acid; all highly substituted with acetic acid). Hemicelluloses from hardwoods are typically high in xylose, while those in softwoods contain more hexoses [3].

Cellulose, in contrast to hemicellulose, is a semicrystalline polymer of pure glucose linked by beta-glucoside bonds. The beta linkages form linear strands that establish extensive H-bonds between them, leading to a highly stable structure that is quite resistant to degradation. Many chemical approaches have been explored to hydrolyze cellulose, though none are completely satisfactory; currently, dilute acid hydrolysis procedures are being proposed for several near-term commercialization efforts until more effective technologies are available [5].

2.1.3. Enzymatic cellulose hydrolysis. Enzymatic hydrolysis by cellulases is the ultimate goal in biomass processing for fermentation: this method has the advantages of reduced sugar loss through side reactions and it is less corrosive of process equipment [18]. In addition, the hydrolyzed product requires no neutralization prior to fermentation [19].

Cellulases consist of multicomponent enzyme complexes acting synergistically: complete cellulose hydrolysis requires the activity of an endoglucanase, which cleaves interior regions of cellulose polymers; an exoglucanase, which cleaves cellobiose units from the ends of cellulose polymers; and a beta-glucosidase, which cleaves cellobiose into its glucose subunits [20]. Because of the complexity and insolubility of the substrate, cellulase catalysis is not only relatively slow, but it is also understood much less completely than other enzymes, despite over four decades of cellulase research [8]. One of the most important organisms in the development of cellulase enzymes is Trichoderma reesei, the "ancestor of many of the most potent enzyme- producing fungi in commercial use today"[3]. By 1979, genetic enhancement had produced mutants with up to 20 times greater cellulose productivity than the original organisms found in World War II; today, surprisingly, the most lucrative cellulase market is in the manufacture of stone-washed jeans [3].

2.1.4. Fermentation. Fermentation of glucose to ethanol is performed by numerous bacteria, yeasts, and other fungi, and several yeasts have also been identified that can convert xylose to ethanol. Pentose fermentation to ethanol does not commonly co-occur with hexose fermentation to ethanol, however, spurring efforts to combine these two fermentation pathways into single organisms. Genetic engineering has since provided both bacteria and yeasts capable of fermenting both 5-carbon and 6-carbon sugars [21, 22].

Although hydrolysis of biomass cellulose by cellulases was once performed as a distinct step between pretreatment and fermentation, fermentation can begin as soon as glucose subunits are released from cellulose. The realization of this led to the development of the Simultaneous Saccharification and Fermentation (or Co-Fermentation) Process (SSF or SSCF), which now provides the great advantage of simultaneous cellulose hydrolysis and glucose fermentation [13]. This process enhances cellulase activity by relieving the product inhibition of beta-glucosidase by glucose, since the products are consumed as soon as they are produced. SSF has been patented by the Gulf Oil Company and the University of Arkansas [3]. With the availability of organisms that can ferment both pentoses and hexoses, all biomass sugars may now be simultaneously fermented in SSF/SSCF processes [23, 24]

Product inhibition of exocellulases is not eliminated, however, unless the glucose dimer cellobiose is also consumed or hydrolyzed. Since conventional S. cerevisiae strains do not metabolize cellobiose, and since cellulase preparations with sufficient beta-glucosidase activity to hydrolyze all the cellobiose are expensive to produce, much research has been directed toward the use of native cellobiose-utilizing yeast strains in SSF, either independently or in co-culture with S. cerevisiae [18].

2.1.5. Fermentation following gasification. A radically different approach to preparing biomass substrates for fermentation is found in biomass gasification. In this process, biomass is converted to synthesis gas, consisting primarily of CO, CO_2, and H_2, in addition to CH4 and N2 (25). After gasification, anaerobic bacteria such as Clostridium ljungdahlii can ferment the CO, CO_2, and H_2 into ethanol by an acetogenic process [26-29]. One advantage of the process is that, unlike acid and enzymatic hydrolysis methods, gasification can convert essentially all of the biomass, including lignin, to syngas that can be potentially fermented by bacteria [30]. Higher rates of fermentation are also achieved because the process is limited by the transfer of gas into the liquid phase instead of the rate of substrate uptake by the bacteria.

2.2. Feedstock Options

Corn and sugar cane are not long-term options for ethanol generation because of their value as foods. Exploration of various non-food forms of biomass, principally wastes, is therefore an active area of research.

Worldwide, rice straw has the greatest quantitative potential for bioethanol production, estimated at 205 gigaliters per year; this potential is concentrated in Asia, which as a region could produce up to 291 gigaliters per year of ethanol from rice straw in combination with wheat straw and corn stover. Europe has the next-largest supply of agricultural wastes, primarily in the form of wheat straw (69.2 gigaliters per year potential ethanol production); followed by North America, in which corn stover forms the majority of agricultural wastes and could supply an estimated 38.4 gigaliters per year of ethanol [11]. Bagasse, or waste derived from sugar cane, is widely available in tropical areas and is being explored by BC International Corporation (BCI), while municipal solid waste has attracted the attention of Masada Resources Group, LLC; these two companies are currently planning construction of unique biomass-to-ethanol plants [31].

Corn stover or fiber, a by-product of the corn wet-milling industry consisting of corn hulls and residual starch, is the subject of great interest as a possible substrate for ethanol production in the United States. Conversion of the starch along with the lignocellulosic components in the corn fiber could increase ethanol yields from a corn wet mill by 13 percent. In a recent study utilizing the bioethanol process development unit at the U.S. National Renewable Energy Laboratory (NREL), corn fiber was used to support continuous, integrated operation of the plant. The fiber was pretreated by high-temperature, dilute sulfuric-acid hydrolysis, and the cellulose was converted to ethanol using simultaneous saccharification and fermentation using a commercially-available cellulase and conventional Saccharomyces cerevisiae yeast that did not utilize 5-carbon sugars. Despite difficulties with bacterial contamination, which are expected to diminish with the use of recombinant, xylose- and arabinose-utilizing fermentative organisms, the attempt was successful and indicates that corn fiber could become a valuable feedstock in the United States [31]. The use of agricultural wastes is not without potential drawbacks, however. Most crop residues are currently plowed into the soil to sustain soil quality by increasing the soil organic carbon pool, enhancing activity of soil fauna, and minimizing soil erosion, and soil scientists caution that diversion of waste biomass for fuel must be undertaken cautiously [32].

Softwood forest thinnings are also being explored as potential feedstocks. Lumber manufacturing, timber harvesting, and thinning of forests to prevent wildfires generate a large quantity of softwood residues that require environmentally sound and cost-effective methods of disposal. Research in this area in the United States is currently focusing on dilute sulfuric acid hydrolysis and SO_2-steam explosion pretreatments, followed by fermentation by a Saccharomyces cerevisiae mutant yeast adapted to the inhibitory extractives and lignin degradation products present in the softwood hydrolysates [33]. Testing of recombinant xylosefermenting yeasts is also planned, and investigation is underway by Kemestrie, Inc. (Sherbrooke, QC, Canada) to identify high-value coproducts that may be derived from softwoods, focusing on antioxidants and other extractives [3].

2.3. Cellulase Engineering

The second important area in which improvement is needed for the commercialization of fuel ethanol is the conversion of lignocellulosic feedstock into the sugars to be fermented. Most current work in this area concentrates on improvement of cellulase expression, activity, and production efficiency, with the goal of reducing the cost and increasing the extent of cellulose hydrolysis.

Cellulase cost is a critical limiting factor in lignocellulose feedstock preparation. Current estimates of cellulase cost range from 3 0–50 cents per gallon of ethanol; a goal of 5 cents per gallon of ethanol is envisioned [3]. Thus, a 10-fold improvement in specific activity, production efficiency, or some combination thereof, is required.

Cellulase improvement in any of the following five critical areas could substantially improve the feasibility of bioethanol commercialization: thermostability, acid tolerance (to withstand pretreatment acidification), cellulose binding affinity, specific activity, and reduced nonspecific binding to lignin [14]. While these features are theoretically approachable by genetic engineering techniques, use of these techniques is presently limited by the incomplete understanding of cellulase catalysis. A primary reason for this is that cellulose-cellulase systems involve soluble enzymes working on insoluble substrates, which represents a substantial increase in complexity from homogeneous enzyme-substrate systems. In addition, the catalytic system involves the synergistic activities of three different enzymes [3]. Still, a number of promising avenues are currently being explored.

2.3.1. Cellulase component engineering. The most fundamental improvements that are needed are within the cellulase components themselves, these are the endoglucanases, exoglucanases, and cellobiohydrolases (or beta-glucosidases). Using Trichoderma, Clostridium, Cellulomonas, and Thermobifida, among others, efforts are underway to improve activity, expression, and specificity of these components through site-directed mutagenesis, use of heterologous promoters to direct transcription, and modeling to reveal structure-function relationships [34–39].

One of the most active cellulase components known is the endoglucanase E1 from Acidothermus cellulolyticus. Two leading industrial enzyme producers, Novozymes (www.novozymes.com) and Genencor International (www.genencor.com), are currently contributing to the cellulase improvement effort with support from DOE. In 1998, J. Sakon and colleagues at the University of Arkansas showed that performance of a ternary system was improved 13 percent by site-directed modification of one active site amino acid in Acidothermus E1; currently they are pursuing E1 mutations that modify the biomass interactive surface.

2.3.2. Chimeric cellulase systems. Cellulase components from diverse organisms, primarily bacteria and fungi, are being combined in ways that yield overall improved activity. Baker and colleagues have successfully combined bacterial and fungal cellulases in vitro [40], showing that these mixtures can be competitive with a native ternary system from T. reesei [41]. Work with expansins, proteins that enable extension of plant cell walls during plant cell growth, has also shown enhancement of hydrolysis of microcrystalline cellulose in a mixed Trichoderma cellulase preparation [42]. The initial approaches to developing artificial cellulase systems, still instructive after nine years, are reviewed in [43].

2.3.3. Heterologous expression. The next logical step in chimeric cellulase systems is the cloning of cellulases from one organism into another; this avenue is being explored as well, as shown by the expression of the T. reesei cellobiohydrolase I in Pichia pastoris [44]. Especially important for commercial production, cellulases are being expressed in plants such as tobacco and potatoes, potentially providing more abundant sources of the enzymes [45].

Another innovative approach to the heterologous expression of cellulases is the expression of heat-activated cellulases within biomass crops themselves, with the idea that the plants grow normally until harvested and exposed to elevated temperatures, at which point heat- activated cellulases hydrolyze the cellulose without need for externally added enzymes [46].

2.3.4. Cellulase performance assays. Convenient, accurate, efficient assays are central to the development of any new technology. The diafiltration saccharification assay (DSA) developed at the NREL produces precise, detailed progress curves for enzymatic saccharification of cellulosic materials under conditions that mimic those of SSF. From this method, it is possible to describe the performance of a given cellulase preparation over a wide range of loading and reaction times with comparatively little data [47, 48, 3].

2.3.5. Proteomic analysis, microarray analysis, and modeling. Proteomics is an emerging set of techniques that has proven extremely useful in understanding the interactions of multienzyme systems. Hydrolysis of complex organic substrates is an ideal candidate for proteomic analysis, as it involves a number of enzymes: β3-1,4-endoglucanases, β3-1,4-cellobiohydrolases, xylanases, β3-glucosidases, α -L-arabinofuranosidase, acetyl xylan esterase, β3- mannanase, and α-glucuronidase in T. reesei, for example. At the NREL, the expression of these enzymes is being investigated under various conditions by proteomic methods and compared to corresponding enzyme activities using the DSA assay [3].

To reveal gene expression responses to environmental conditions in both wild-type and genetically engineered microbes, microarray analysis is underway and could become a valuable industrial tool for evaluation of new recombinant organisms [49]. Mathematical molecular analysis is also being employed to gain greater understanding of structure-function relationships to complement the physiological understanding provided by proteomic and microarray analysis. Current work includes molecular mechanics efforts by Brady and colleagues at Cornell University as well as Palma and colleagues; cellulase crystallization work is also in progress by a number of groups [50-53].

2.4. Fermentation Technologies

While fermentation of glucose into ethanol is a well-understood process that occurs widely among microorganisms, the fermentation of pentoses such as xylose, which are abundant in biomass, has posed significant challenges. Recently, however, this challenge has been addressed by creating recombinant yeasts and bacteria [21, 22], although the solution may not yet have been fully optimized.

A second important aspect limiting commercialization is the fermentation efficiency of the microorganisms: in typical fermentation pathways for glucose and xylose to ethanol, one contributor to the high cost of ethanol production is the loss of half of the fixed carbon to products other than ethanol [5].

The ideal bioethanol-fermenting microorganism would therefore readily ferment all biomass sugars, resist toxic effects of aromatic lignin subunits and other inhibitory byproducts such as acetate, be thermostable and acid-tolerant, and produce a highly active cellulase multienzyme complex [54].

2.4.1. Pentose fermentation. As mentioned above, Escherichia, Klebsiella, and Zymomonas have now been engineered to ferment not only glucose but also xylose and arabinose sugars [5, 54, 55]. Some of these are already experiencing commercial use as well: BC International Corporation (www.bcintl-corp.com) is using genetically engineered Escherichia coli to produce ethanol from biomass sugars, and Arkenol Inc. (www.arkenol.com) is using Zymomonas in its concentrated-acid process.

In another example, Zymomonas mobilis has been transformed with Escherichia coli xylose isomerase, xylulokinase, transaldolase, and transketolase genes. Expression of the added genes are under the control of Zymomonas mobilis promoters. This genetically modified microorganism, patented by the Midwest Research Institute, is now able to ferment mixtures of xylose, arabinose, and glucose to produce ethanol [56, 57].

2.4.2. Combined cellulolysis and fermentation. Consolidated bioprocessing (CBP), in which the production of cellulolytic enzymes, hydrolysis of biomass, and fermentation of resulting sugars to desired products occur in one step, is currently envisioned as the most promising and eminently achievable path toward optimally efficient bioethanol production [54]. Efforts to develop such a culture through engineering fermentative capacity into cellulolytic organisms, as well as the alternative, engineering cellulolytic capacity into fermentative organisms, are both underway and have been reviewed extensively [54].

In one example, Ingram and colleagues cloned two Erwinia endoglucanase genes into an ethanol-producing Klebsiella species, producing a new microbe that produced up to 22 percent more ethanol when fermenting crystalline cellulose synergistically with added fungal cellulases [58]. Cellulase genes have also been introduced into Lactobacillus, although not necessarily for biomass utilization [59], and cellobiose utilization capability has been engineered into Saccharomyces cerevisiae [60]. In an alternative example, with the additional goal of offering improved relief from product inhibition in SSF, in which cellobiose inhibition of exoglucanase is problematic, ethanol-producing genes have been successfully introduced into native cellobioseutilizing bacteria [61, 62].

2.4.3. Synergistic co-cultures. In experiments involving the cellobio se-fermenting recombinant, Klebsiella oxytoca P2, in co-cultures with ethanol-tolerant strains of Saccharomyces pastorianus, Kluyveromyces marxianus, and Zymomonas mobilis, the combinations produced more ethanol, more rapidly, than any of the constituent strains. This was accomplished by early ethanol production by K. oxytoca, while ethanol produced in the later stages was primarily by the more ethanol-tolerant strain [18].

2.4.4. Improved thermotolerance. Ethanol fermentation at elevated temperatures (>55°C) would facilitate product recovery, but thermophilic bacteria are poor ethanol producers. In addition, thermophilic Clostridium and Thermoanaerobium species have been investigated for potential as ethanol producers, but were consistently limited by end-product inhibition and solvent-induced membrane damage [63].

In addition, efforts are underway to eliminate acid production during fermentation through genetic engineering, enabling use of salt-intolerant thermophilic strains like Thermoanaerobacterium thermosaccharolyticum, a microbe tolerant to high levels of ethanol but intolerant of salt accumulation during pH-controlled fermentations. If such a thermotolerant organism could be improved further to produce high-activity cellulases, a highly productive, anaerobic, ethanol-producing strain could result. Cellulase production could, however, pose an insurmountable energy burden to a fermentative organism; the energetic considerations of this combination are being evaluated [5].

2.4.5. Fermentation of synthesis gas. Rajagopalan and coworkers report the discovery of a clostridial bacterium, P7, that converts mixtures of CO, CO_2, and N2 into ethanol, butanol and acetic acid, with high ethanol production and selectivity compared to previous isolates. The authors report process parameters and consider options for improving ethanol yield [64].

2.5. Coproduct Development

Finally, the investigation of potential ethanol coproducts is underway. Biomass sugars can support the production of many other products along with ethanol, including organic acids and other organic alcohols, 1 ,2-propanediol, and aromatic chemical intermediates. If these coproducts were sufficiently valuable, they could help greatly offset costs of ethanol production. However, such coproducts must be chosen carefully to ensure that sufficient markets are available [65].

Additional coproducts may be available from lignin: this material is present at 15–30 percent by weight in all lignocellulosic biomass, and any bioethanol production process will have lignin as a residue. A team of researchers from the NREL, the University of Utah, and Sandia National Laboratories is working to develop a process for making oxygenate fuel additives from lignin; these processes are chemical in nature and are detailed in other materials referenced [66- 68].

3. Research Priorities

The consensus among researchers and supporters of bioethanol research, in addition to those engaged in commercial projects, is that the improvement of cellulase enzyme activity and cellulase production, both to increase the efficiency of release of fermentable sugars from biomass and to reduce cellulase cost, are two of the greatest advances needed in the effort to commercialize fuel ethanol production [19, 5]. In addition is the development of enzymatic pretreatment processes to release lignin from carbohydrate components [42, 9] and further improvement of fermentative organisms [69, 64], with the particular goal of designing microbes capable of consolidated bioprocessing [54].

4. Commercialization

4.1. Cellulases

Although many commercial preparations of cellulase exist, costs have remained high because present applications are in higher-value markets (food and clothing) than fuels. In addition, these applications typically require much less than 100 percent cellulose hydrolysis, in contrast to ethanol production; much improvement is therefore needed to advance the current cellulase enzyme industry to the point at which it can support the fuel ethanol industry. As a result, many fuel ethanol commercialization efforts are choosing to use acid hydrolysis techniques for cellulose hydrolysis until cellulase preparations become less expensive or until recombinant microbes capable of combined cellulolysis and fermentation are perfected [3]. Toward this goal, the DOE biofuels program is working with the two largest global enzyme producers, Genencor International and Novozymes Biotech Incorporated, to achieve a 10-fold reduction in cost of cellulases [3]

4.2. Ethanol Plants

The first dedicated large-scale plants for the conversion of waste biomass to ethanol are now in planning and/or construction phases by BCI and the Masada Resource Group (www.masada.com), while Iogen Corporation (www.iogen.com) is currently operating a 50 ton per week pilot plant. BCI and the DOE Office of Fuels Development have formed a cost-shared partnership to develop a biomass-to-ethanol plant intended to produce 20 million liters of ethanol per year initially from sugar cane waste (bagasse) and other biomass, utilizing an existing ethanol plant in Jennings, LA. Dilute acid hydrolysis will be used to recover sugar from bagasse initially, allowing for addition of enzyme hydrolysis when cellulases become less expensive. A proprietary genetically-engineered microbe will ferment the sugars to ethanol. BCI is also planning to operate a plant in Gridley, CA, in which cellulases will be used in conversion of commercial rice straw to ethanol, again with partial DOE support.

The Masada plant is expected to produce 9.5 million gallons from municipal solid waste using Masada's patented CES OxyNolTM concentrated acid hydrolysis technology in New York [3].

Petro-Canada, the second largest petroleum refining company in Canada, began to co-fund research and development on biomass-to-ethanol technology with Iogen in 1997. Petro-Canada, Iogen, and the Canadian government then began plans to fund construction of a demonstration plant based on Iogen's cellulase enzyme technology in an SSF process [3]. The plant of Iogen, a leading producer of cellulases, has completed a 40 ton per day biomass-to-ethanol demonstration facility that is now in its start-up phase [5].

In the pulp and paper industry, Tembec and Georgia Pacific are using dilute acid hydrolysis to dissolve hemicellulose and lignin from wood, producing a cellulose pulp that can be fermented to ethanol. The lignin is then used to generate energy, through combustion, or converted to other products such as concrete additives and soil stabilizers [3].

Pursuing the gasification and syngas-to-ethanol fermentation, BioEngineering Resources, Inc. (BRI) has developed syngas technology to the extent that plans are underway to pilot the technology as a first step toward commercialization. BRI has developed bioreactor systems for fermentation that result in retention times of minutes or less, yielding low equipment costs [3].

REFERENCES

[1] Turner, J. (1999). *A realizable renewable energy future*, Science 285:687-689.

[2] United States Department of Energy (2003). *Biofuels for Sustainable Transportation.* Office of Energy Efficiency and Renewable Energy, http://www.ott.doe.gov/biofuels/advanced _bioethanol.html.

[3] United States Department of Energy (2003). *Advanced Bioethanol Technology*, Office of Energy Efficiency and Renewable Energy, http://www.ott.doe.gov/biofuels/advanced bioethanol.html.

[4] Kaltschmitt, M., G. Reinhardt, and T. Stelzer (1997). *Life cycle analysis of biofuels under different environmental aspects*, Biomass Bioenergy 12:121-134.

[5] Mielenz (2001). *Ethanol production from biomass: technology and commercialization status*, Current Opinion in Microbiology 4:324-329.

[6] United States Department of Energy (2004). *Energy Security and the U.S. Transportation Sector*, Office of Energy Efficiency and Renewable Energy, Biomass Program, http://www.eere.energy.gov/biomass/national_energy_security.html#security.

[7] United States Department of Energy (2004). *Alternative Fuels Data Center*, Office of Energy Efficiency and Renewable Energy, Clean Cities Program, http://www.eere.energy.gov/cleancities/afdc/infrastructure/station_counts.html.

[8] Sheehan, J. (2001). *The road to bioethanol: a strategic perspective of the US Department of Energy's national ethanol program,* in Glycosyl Hydrolases for Biomass Conversion (M. Himmel, J. O. Baker, and J. Saddler, Eds.). American Chemical Society, Washington, D. C., pp 2-25.

[9] Sun, Y., and J. Cheng (2002). *Hydrolysis of lignocellulosic materials for ethanol production: a review*, Bioresource Technol 83:1-11.

[10] Goldemburg, J., S. Coelho, P. Nastari, and O. Lucon (2004). *Ethanol learning curve -- the Brazilian experience*, Biomass Bioenergy 26:301-304.

[11] Kim, S., and B. Dale (2004). *Global potential bioethanol production from wasted crops and crop residues*, Biomass Bioenergy 26:361-375.

[12] Himmel, M. E., S. A. Adney, J. O. Baker, R. Elander, J. D. McMillan, R. A. Nieves, J. J. Sheehan, S. R. Thomas, T. B. Vinzant, and M. Zhang (1997). *Advanced Bioethanol Production Technologies: A Perspective, in* Fuels and Chemicals from Biomass. American Chemical Society, Washington, DC, pp 1-45.

[13] Wyman, C. E. (1999). *Biomass ethanol: Technical progress, opportunities, and commercial challenges*, Annu Rev Energy Environ 24:189-226.

[14] Wooley, R., M. Ruth, D. Glassner, and J. Sheehan (1999). *Process design and costing of bioethanol technology: A tool for determining the status and direction of research and development*, Biotechnol Prog 15:794-803.

[15] Mosier, N., C. E. Wyman, B. Dale, R. Elander, Y. Y. Lee, M. Holzapple, and M. Ladisch (2005). *Features of promising technologies for pretreatment of lignocellulosic biomass*, Bioresource Technol 96:673-686.

[16] Amen-Chen, C., H. Pakdel, and C. Roy (2001). *Production of monomeric phenols by thermochemical conversion of biomass: A review*, Bioresource Technol 79:277-299.

[17] United States Department of Energy (2004). Pretreatment Technology Evaluation, Office of Energy Efficiency and Renewable Energy, Biomass Program, http://www.eere.energy.gov/biomass/technology_evaluation.html.

[18] Golias, H., G. J. Dumsday, G. A. Stanley, and N. B. Pamment (2002). *Evaluation of a recombinant Klebsiella oxytoca strain for ethanol production from cellulose by simultaneous saccharification and fermentation: comparison with native cellobiose-utilising yeast strains and performance in co-culture with therm otolerant yeast and Zymomonas mobilis,* J Biotechnol 96:155-168.

[19] Himmel, M. E., J. O. Baker, and J. N. Saddler (2001). *Glycosyl hydrolases for biomass conversion.* American Chemical Society : Distributed by Oxford University Press, Washington, DC.

[20] Himmel, M. E., W. S. Adney, J. O. Baker, R. A. Nieves, and S. R. Thomas (1996). *Cellulases: Structure, Function, and Application, in* Handbook on Bioethanol (C. E. Wyman, Ed.). Taylor & Francis, Washington, DC, pp 143-161.

[21] Dien, B., R. Hespell, H. Wyckoff, and R. Bothast (1998). *Fermentation of hexose and pentose sugars using a novel ethanologenic* Escherichia coli strain, Enzyme Microb Technol 23:366-371.

[22] Toivari, M., A. Aristidou, L. Ruohonen, and M. Penttila (2001). *Conversion of xylose to ethanol by recombinant* Saccharomyces cerevisiae*: Importance of xylulokinase (XKS1) and oxygen availability*, Metabolic Eng 3:236-249.

[23] McMillan, J. D., M. M. Newman, D. W. Templeton, and A. Mohagheghi (1999). *Simultaneous saccharification and cofermentation of dilute-acid pretreated yellow poplar hardwood to ethanol using xylose-fermenting* Zymomonas mobilis, Appl Biochem Biotechnol 77/79:649-665.

[24] Teixeira, L. C., J. C. Linden, and H. A. Schroeder (2000). *Simultaneous saccharification and cofermentation of peracetic acid-pretreated biomass*, Appl Biochem Biotechnol 84/86:111-127.

[25] McKendry, P. (2002). *Energy production from biomass. Part 3: Gasification technologies*, Bioresource Technology 83:55-63.

[26] Ljungdahl, L. G. (1986). *The autotrophic pathway of acetate synthesis in acetogenic bacteria*, Annu Rev Microbiol 40:415-450.

[27] Wood, H. G., S. W. Ragsdale, and E. Pezacka (1986). *The acetyl-CoA pathway of autotrophic growth*, FEMS Microbiol Rev 39:345-362.

[28] Abrini, A., H. Naveau, and E. J. Nyns (1994). Clostridium autoethanogenum*, sp. nov., an anaerobic bacterium that produces ethanol from carbon monoxide*, Arch Microbiol 161 :345-351.

[29] Phillips, J. R., E. C. Clausen, and J. L. Gaddy (1994). *Synthesis gas as substrate for the biological production of fuels and chemicals*, Appl Biochem Biotechnol 45/46:145-154.

[30] Reed, T. B., and D. Jantzen (1979). *Biomass gasification: Principles and technology*, Energy Technology Review 67:27-90.

[31] Schell, D. J., C. J. Riley, N. Dowe, J. Farmer, K. N. Ibsen, M. F. Ruth, S. T. Toon, and R. E. Lumpkin (2004). *A bioethanol process development unit: initial operating experiences and results with a corn fiber feedstock,* Bioresource Technol 91:179-188.

[32] Lal, R. (2005). *World crop residues production and implications of its use as a biofuel,* Environ Int 3 1:575-584.

[33] Keller, F. A., D. Bates, R. Ruiz, and Q. Nguyen (1998). *Yeast adaptation on softwood prehydrolysate*, Appl Biochem Biotechnol 70:137-148.

[34] Boer, H., T.T. Teeri, and A. Koivula (2000). *Characterization of* Trichoderma reesei *cellobiohydrolase Cel7A secreted from* Pichia pastoris *using two different promoters*, Biotechnol Bioeng 69:486-494.

[35] Boisset, C., C. Fraschini, M. Schulein, B. Henrissat, and H. Chanzy (2000). *Imaging the enzymatic digestion of bacterial cellulose ribbons reveals the endo character of the cellobiohydrolase Cel6A from* Humicola insolens *and its mode of synergy with cellobiohydrolase Cel7A*, Appl Environ Microbiol 66:1444-1452.

[36] Brun, E., P.E. Johnson, A.L. Creagh, P. Tomme, P. Webster, C.A. Haynes, and L.P. McIntosh (2000). *Structure and binding specificity of the second N-terminal cellulose-binding domain from* Cellulomonas fimi *endoglucanase C*, Biochem 39:2445-2458.

[37] Irwin, D. C., S. Zhang, and D. B. Wilson (2000). *Cloning, expression and characterization of a family 48 exocellulase, Cel48A, from* Thermobifida fusca, Eur J Biochem 267:4988-4997.

[38] Quixley, K. W., Reid S.J. (2000). *Construction of a reporter gene vector for* Clostridium beijerinckii *using a Clostridium endoglucanase gene*, J Mol Microbiol Biotechnol 2:53- 57.

[39] Zhang, S., D. C. Irwin, and D. B. Wilson (2000). *Site-directed mutation of noncatalytic residues of* Thermobifida fusca *exocellulase Cel6B*, Eur J Biochem 267:3101-3115.

[40] Baker, J. O., W. S. Adney, R. A. Nieves, S. R. Thomas, and M. E. Himmel (1995). *Synergism in Binary Mixtures of Bacterial and Fungal Cellulases: Endo/Exo, Exo/Exo, and Endo/Endo Interactions, in* Enzymatic Degradation of Insoluble Polysaccharides (J.

[41] N. Saddler and M. H. Penner, Eds.). American Chemical Society, Washington, DC, pp 113- 141.

[42] Baker, J. O., C. I. Ehrman, W. S. Adney, S. R. Thomas, and M. E. Himmel (1998). *Hydrolysis of cellulose using ternary mixtures of purified cellulases*, Appl Biochem Biotechnol 70:395-403.

[43] Baker, J. O., M. R. King, W. S. Adney, S. R. Decker, T. B. Vinzant, S. L. Lantz, R. E. Nieves, S. R. Thomas, L. -C. Li, D. J. Cosgrove, and M. E. Himmel (2000). *Investigation of the cell wall loosening protein expansion as a possible additive in the enzymatic saccharification of lignocellulosic biomass*, Appl Biochem Biotechnol. 84-86:217-223.

[44] Thomas, S. R., R. A. Laymon, Y. C. Chou, M. P. Tucker, T. B. Vinzant, W. S. Adney, J. O. Baker, R. A. Nieves, J. R. Mielenz, and M. E. Himmel (1995). *Initial Approaches to Artificial Cellulase Systems for Conversion of Biomass to Ethanol, in* Enzymatic Degradation of Insoluble Polysaccharides (J. N. Saddler and M. H. Pennery, Eds.). American Chemical Society, Washington, DC, pp 208-23 6.

[45] Godbole, S., S. R. Decker, R. A. Nieves, W. S. Adney, T. B. Vinzant, J. O. Baker, S. R. Thomas, and M. E. Himmel (1999). *Cloning and expression of* Trichoderma reesei *CBH I in Pichia pastoris*, Biotechnology Progress 15:828-833.

[46] Dai, Z., B. S. Hooker, D. B. Anderson, and S. R. Thomas (2000). *Expression of* Acidothermus cellulolyticus *endoglucanase E1 in transgenic tobacco: Biochemical characteristics and physiological effects*, Transgenic Research 9:43-54.

[47] Office of News Services (2001). New CU-Boulder Research May Reduce Renewable Fuels Costs, B. University of Colorado, http://www.colorado.edu/PublicRelations/NewsReleases/2001/1 244.html.

[48] Baker, J. O., T. B. Vinzant, C. I. Ehrman, W. S. Adney, and M. E. Himmel (1997) *A membrane-reactor saccharification assay to evaluate the performance of cellulases under simulated SSF conditions*, Appl Biochem Biotechnol 63:585-595.

[49] Vinzant, T. B., C. I. Ehrman, and M. E. Himmel (1997). *SSF of pretreated hardwoods. Effect of native lignin content*, Appl Biochem Biotechnol 62:97-102.

[50] Tao, H., R. Gonzalez, A. Martinez, M. Rodriguez, L. Ingram, J. Preston, and K. Shanmugam (2001). *Engineering a homoethanol pathway in* Escherichia coli. *Increased glycolytic flux and levels of expression of glycolytic genes during xylose fermentation*, J Bacteriol 185:2979-2988.

[51] Sakon, J., W. Adney, M. Himmel, S. Thomas, and P. Karplus (1996). *Crystal structure of thermostable Family 5 endocellulase EI from* Acidothermus cellulolyticus *in complex with cellotetraose*, Biochemistry 35:10648-10660.

[52] Shirai, T., H. Ishida, J. Noda, T. Yamane, K. Ozaki, Y. Hakamada, and S. Ito (2001). *Crystal structure of alkaline cellulase K. Insight into the alkaline adaptation of an industrial enzyme*, J Mol Biol 3 10:1079-1087.

[53] Hirvonen, M., and A. C. Papageorgiou (2003). *Crystal structure of a family 45 endoglucanase from* Melanocarpus albomyces. *Mechanistic implications based on the free and cellobiose-bound forms*, J Mol Biol 329:403 -410.

[54] Mandelman, D., A. Belaich, J. P. Belaich, N. Aghajari, H. Driguez, and R. Haser (2003). *X-Ray crystal structure of the multidomain endoglucanase Cel9G from* Clostridium cellulolyticum *complexed with natural and synthetic cello-oligosaccharides*, J Bacteriol 185:4127–4135.

[55] Lynd, L. R., P. J. Weimer, W. H. van Zyl, and I. S. Pretorius (2002). *Microbial cellulose utilization. Fundamentals and biotechnology*, Microbiol Molec Biol Rev 66:506-577.

[56] Dien, B. S., M. A. Cotta, and T. W. Jeffries (2003). *Bacteria engineered for fuel ethanol production. Current status*, Appl Microbiol Biotechnol 63:258-266.

[57] Picataggio, S. K., M. Zhang, C. K. Eddy, K. A. Deanda, and M. Finkelstein (1996). *Recombinant Zymomonas for pentose fermentation,* U.S. Patent 5514583.

[58] Zhang, M., Y.-C. Chou, S. K. Picataggio, and M. Finkelstein (1998). *Single* Zymomonas mobilis *strain for xylose and arabinose fermentation,* U.S. Patent 5843760.

[59] Zhou, S., F. C. Davis, and L. O. Ingram (2001). *Gene integration and expression and extracellular secretion of* Erwinia chrysanthemi *endoglucanase CelY (celY) and CelZ (celZ) in ethanologenic* Klebsiella oxytoca *P2*, Appl Environ Microbiol 67:6-14.

[60] Cho, J., Y. Choi, and D. Chung (2000). *Expression of* Clostridium thermocellum *endoglucanase gene in* Lactobacillus gasseri *and* Lactobacillus johnsonii *and characterization of the genetically modified probiotic lactobacilli*, Curr Microbiol 40:257-263.

[61] McBride, J. E., J. J. Zietsman, W. H. van Zyl, and L. R. Lynd (2005). *Utilization of cellobiose by recombinant ß-glucosidase-expressing strains of Saccharomyces cerevisiae: Characterization and evaluation of the sufficiency of expression,* Enzyme Microb Technol 37:93-101.

[62] Wood, B. E., and L. O. Ingram (1992). *Ethanol production from cellobiose, amorphous cellulose, and crystalline cellulose by recombinant* Klebsiella oxytoca *containing*

chromosomally integrated Zymomonas mobilis *genes for ethanol production and plasmids expressing thermostable cellulase genes from* Clostridium thermocellum, Appl Environ Microbiol 58:2103-2110.

[63] Moniruzzaman, M., X. Lai, S. W. York, and L. O. Ingram (1997). *Extracellular melibiose and fructose are intermediates in raffinose catabolism during fermentation to ethanol by engineered enteric bacteria*, J Bacteriol 179:1880-1886.

[64] Ingram, L. (1990). *Ethanol tolerance in bacteria*, CRC Crit Rev Biotechnol 9:305-319.

[65] Rajagopalan, S., R. P. Datar, and R. S. Lewis (2002). *Formation of ethanol from carbon monoxide via a new microbial catalyst*, Biomass and Bioenergy 23 :487-493.

[66] Lynd, L., C. Wyman, and T. Gerngross (1999). *Biocommodity engineering*, Biotechnol Prog 15:777-793.

[67] Miller, J. E., L. Evans, A. Littlewolf, and D. E. Trudell (1999). *Batch microreactor studies of lignin and lignin model compound depolymerization by bases in alcohol solvents*, Fuel 78:1363-1366.

[68] Shabtai, J., and et al. (1999). *Process for conversion of lignin to reformulated hydrocarbon gasoline,* U.S. Patent 5,959,167.

[69] Shabtai, J. S., W. Zmierczak, E. Chornet, and D. K. Johnson (1999). *Lignin conversion to high-octane fuel additives,* Fourth Biomass Conference of the Americas, Oakland, CA.

[70] Wyman, C. E. (2001). *Twenty years of trials, tribulations, and research progress in bioethanol technology: Selected key events along the way*, Appl Biochem Biotechnol 91-93 :5-21.

In: Bioethanol: Production, Benefits and Economics
Editor: Jason B. Erbaum
ISBN 978-1-60741-697-5

Chapter 10

ADVANCED BIOETHANOL TECHNOLOGY*

United States Department of Energy

Current ethanol production is based on corn grain or other starch or sugar sources which make up only a very small portion of plant material. With advanced bioethanol technology, ethanol can also be made from cellulose and hemicellulose (the components that give plants their structure), which make up the bulk of plant material.

Potential feedstocks for advanced bioethanol technology include corn stover (stalks and husks) and other agricultural residues, wood chips and other forestry residues, paper and other municipal wastes, food processing and other plant-derived industrial wastes, and dedicated energy crops of fast-growing trees or grasses.

Advanced technology bioethanol would supplement rather than replace grain ethanol, but the huge volume of inexpensive available feedstocks offers potential to greatly expand ethanol production and its economic and environmental benefits.

The U.S. Department of Energy National Biofuels Program is supporting research and development to lower the cost of advanced bioethanol technology, so as to make it a marketplace reality, and has set a goal to have commercial demonstration plants using agricultural residues in operation by 2005.

BIODIESEL

As with ethanol, biodiesel is primarily used as a pollution-reducing additive to conventional diesel, usually in a 20% blend—B20.

U.S. biodiesel production is roughly equally split between soybean oil and recycled restaurant cooking grease—giving biodiesel a reputation for a pleasant "french fry" smell.

2001 biodiesel production was predominantly from soybeans because of a USDA program supporting commodity purchases for increase biofuels production. Both soybean oil and recycled grease are in surplus and biodiesel production uses only a small fraction, so there are no resource constraints.

* Extracted from http://www.eere.energy.gov/states/alternatives/

One primary market for biodiesel is state and federal fleets complying with alternative fuel requirements (EPAct does not cover heavy vehicles, but B20 use—which unlike other alternative fuels requires no new vehicle purchase—gains credits against required light-duty vehicle purchase requirements).

A second primary market is fleets such as city or school bus fleets for which biodiesel makes a visible statement of concern for air quality and customer health.

Biodiesel popularity is growing rapidly with U.S. sales growing from 7 million gallons in 2000 to more than 20 million gallons in 2001. The industry expects to be able to expand capacity rapidly to meet increased demand.

Biodiesel use emits only about half as much carbon monoxide, hydrocarbons, and particulates as petroleum diesel. The cancer-risk contribution of diesel is cut by about 90%. Benefits from B20 use are approximately proportional.

B20 typically costs 8 to 20 cents more than regular diesel.

Ethanol

Ethanol, also known as ethyl alcohol or grain alcohol, can be used either as an alternative fuel or as an octane-boosting, pollution-reducing additive to gasoline. The U.S. ethanol industry produced more than 3.4 billion gallons in 2004, up from 2.8 billion gallons in 2003 and 2.1 billion gallons in 2002. (Renewable Fuels Association and Renewable Fuels Association Ethanol Industry Outlook 2005). Although this number is small when compared with fossil fuel consumption for transportation, as individual states continue to ban the use of MTBE (Methyl Tertiary Butyl Ether) and with the possibility of a Federal ban, ethanol consumption is due for a significant boost. Because of the increased demand on ethanol as a gasoline additive, efforts to increase supplies are necessary in order to meet the increase in demand. As of the start of 2005, 81 ethanol plants in 20 states have the capacity to produce nearly 4.4 billion gallons annually and an additional 16 plants are under construction to add another 750 million gallons of capacity (RFA).

There are four basic steps in converting biomass to bioethanol:

1. Producing biomass results in the fixing of atmospheric carbon dioxide into organic carbon.
2. Converting this biomass to a useable fermentation feedstock (typically some form of sugar) can be achieved using a variety of different process technologies. These processes for fermentation feedstock production constitute the critical differences among all of the bioethanol technology options.
3. Fermenting the biomass intermediates using biocatalysts (microorganisms including yeast and bacteria) to produce ethanol in a relatively dilute aqueous solution is probably the oldest form of biotechnology developed by humankind.
4. Processing the fermentation product yields fuel-grade ethanol and byproducts that can be used to produce other fuels, chemicals, heat and/or electricity.

Corn and other starches and sugars are only a small fraction of biomass that can be used to make ethanol. Advanced Bioethanol Technology allows fuel ethanol to be made from cellulosic (plant fiber) biomass, such as agricultural forestry residues, industrial waste, material in municipal solid waste, trees, and grasses. Cellulose and hemicellulose, the two main components of plants-and the ones that give plants their structure-are also made of sugars, but those sugars are tied together in long chains. Advanced bioethanol technology can break those chains down into their component sugars and then ferment them to make ethanol. This technology turns ordinary low-value plant materials such as corn stalks, sawdust, or waste paper into fuel ethanol. Not quite lead into gold, but maybe more valuable for the U.S. economy, for cutting air pollution, and for reducing dependence on foreign oil.

In: Bioethanol: Production, Benefits and Economics
Editor: Jason B. Erbaum
ISBN 978-1-60741-697-5

Chapter 11

FUEL ETHANOL: BACKGROUND AND PUBLIC POLICY ISSUES*

Brent D. Yacobucci and Jasper Womach

ABSTRACT

In light of a changing regulatory environment, concern has arisen regarding the future prospects for ethanol as a motor fuel. Ethanol is produced from biomass (mainly corn) and is mixed with gasoline to produce cleaner-burning fuel called "gasohol" or "E10."

The market for fuel ethanol, which consumes 6% of the nation's corn crop, is heavily dependent on federal subsidies and regulations. A major impetus to the use of fuel ethanol has been the exemption that it receives from the motor fuels excise tax. Ethanol is expensive relative to gasoline, but it is subject to a federal tax exemption of 5.4 cents per gallon of gasohol (or 54 cents per gallon of pure ethanol). This exemption brings the cost of pure ethanol, which is about double that of conventional gasoline and other oxygenates, within reach of the cost of competitive substances. In addition, there are other incentives such as a small ethanol producers tax credit. It has been argued that the fuel ethanol industry could scarcely survive without these incentives.

The Clean Air Act requires that ethanol or another oxygenate be mixed with gasoline in areas with excessive carbon monoxide or ozone pollution. The resulting fuels are called oxygenated gasoline (oxyfuel) and reformulated gasoline (RFG), respectively. Using oxygenates, vehicle emissions of volatile organic compounds (VOCs) have been reduced by 17%, and toxic emissions have been reduced by approximately 30%. However, there has been a push to change the oxygenate requirements for two reasons. *First*, methyl tertiary butyl ether (MTBE), the most common oxygenate, has been found to contaminate groundwater. *Second*, the characteristics of ethanol-blended RFG-along with high crude oil prices and supply disruptions-led to high Midwest gasoline prices in Summer 2000, especially in Chicago and Milwaukee. Uncertainties about future oxygenate requirements, as both federal and state governments consider changes, have raised concerns among farm and fuel ethanol industry groups and have prompted renewed congressional interest in the substance. Without the current regulatory requirements and incentives, or something comparable, much of ethanol's market would

* Excerpted from CRS Report RL30369 dated Updated March 22, 2000.

likely disappear. Expected changes to the reformulated gasoline requirements could either help or hurt the prospects for fuel ethanol (subsequently affecting the corn market), depending on the regulatory and legislative specifics. As a result, significant efforts have been launched by farm interests, the makers of fuel ethanol, agricultural states, and the manufacturers of petroleum products to shape regulatory policy and legislation.

This report provides background concerning various aspects of fuel ethanol, and a discussion of the current related policy issues.

INTRODUCTION

Ethanol (ethyl alcohol) is an alcohol made by fermenting and distilling simple sugars. Ethyl alcohol is in alcoholic beverages and it is denatured (made unfit for human consumption) when used for fuel or industrial purposes.[1] The biggest use of fuel ethanol in the United States is as an additive in gasoline. It serves as an as an oxygenate (to prevent air pollution from carbon monoxide and ozone), as an octane booster (to prevent early ignition, or "engine knock"), and as an extender of gasoline. In purer forms, it can also be used as an alternative to gasoline in automobiles designed for its use. It is produced and consumed mostly in the Midwest, where corn—the main feedstock for ethanol production--is produced.

The initial stimulus to ethanol production in the mid-1970s was the drive to develop alternative and renewable supplies of energy in response to the oil embargoes of 1973 and 1979. Production of fuel ethanol has been encouraged by a partial exemption from the motor fuels excise tax. Another impetus to fuel ethanol production has come from corn producers anxious to expand the market for their crop. More recently the use of fuel ethanol has been stimulated by the Clean Air Act Amendments of 1990, which require oxygenated or reformulated gasoline to reduce emissions of carbon monoxide (CO) and volatile organic compounds (VOCs).

While oxygenates reduce CO and VOC emissions, they also can lead to higher emissions of nitrogen oxides, precursors to ozone formation. While reformulated gasoline has succeeded in reducing ground-level ozone, the overall effect of oxygenates on ozone formation has been questioned. Furthermore, ethanol's main competitor in oxygenated fuels, methyl tertiary butyl ether (MTBE), has been found to contaminate groundwater. This has led to a push to ban MTBE, or eliminate the oxygenate requirements altogether. High summer gasoline prices in the Midwest, especially in Chicago and Milwaukee, where oxygenates are required, have added to the push to remove the oxygenate requirements. The trade-offs between air quality, water quality, and consumer price have sparked congressional debate on these requirements. In addition, there has been a long-running debate over the tax incentives that ethanol-blended fuels receive.

Fuel ethanol is used mainly as a low concentrate blend in gasoline, but can also be used in purer forms as an alternative to gasoline. In 1999, 99.8% of fuel ethanol consumed in the United States was in the form of "gasohol" or "E10" (blends of gasoline with up to 10% ethanol).[2]

Fuel ethanol is produced from the distillation of fermented simple sugars (e.g. glucose) derived primarily from corn, but also from wheat, potatoes and other vegetables, as well as from cellulosic waste such as rice straw and sugar cane (bagasse). The alcohol in fuel ethanol is identical to ethanol used for other purposes, but is treated (denatured) with gasoline to make it unfit for human consumption.

Ethanol and the Agricultural Economy

Corn constitutes about 90% of the feedstock for ethanol production in the United States. The other 10% is largely grain sorghum, along with some barley, wheat, cheese whey and potatoes. Corn is used because it is a relatively low cost source of starch that can be converted to simple sugars, fermented and distilled. It is estimated by the U. S. Department of Agriculture (USDA) that about 615 million bushels of corn will be used to produce about 1.5 billion gallons of fuel ethanol during the 2000/2001 corn marketing year.[3] This is 6.17% of the projected 9.755 billion bushels of corn utilization [4].

Producers of corn, along with other major crops, receive farm income support and price support. Farms with a history of corn production will receive "production flexibility contract payments" of about $1.186 billion during the 2000/2001 corn marketing year. Emergency economic assistance (P.L. 106-224) more than double the corn contract payments. Corn producers also are guaranteed a minimum national average price of $1.89/bushel under the nonrecourse marketing assistance loan program.[5]

The added demand for corn created by fuel ethanol raises the market price for corn above what it would be otherwise. Economists estimate that when supplies are large, the use of an additional 100 million bushels of corn raises the price by about 4¢ per bushel. When supplies are low, the price impact is greater. The ethanol market is particularly welcome now, when the average price received by farmers is forecast by USDA to average about $1.80 per bushel for the 2000/01 marketing year. This price would be the lowest season average since 1986.

Table 1. Corn Utilization, 2000/2001 Forecast

Quantity(million bushels)	Share of Total Use	
Livestock feed and residual	5,775	59.2%
Food, seed and industrial:	1,980	19.9%
Fuel alcohol	615	6.2%
High fructose corn syrup	550	5.5%
Glucose and dextrose	220	2.2%
Starch	225	2.6%
Cereals and other products	190	1.9%
Beverage alcohol	130	1.3%
Seed	20	0.2%
Exports	2000	20.1%
TOTAL USE	9,775	100.00%
TOTAL PRODUCTION	9,968	

Source: Basic data are from USDA, Economic Research Service, Feed Outlook, March 10, 2000.

The ethanol market of 615 million bushels of corn, assuming a price impact of about 25¢ per bushel on all corn sales, means a possible $2.4 billion in additional sales revenue to corn farmers. In the absence of the ethanol market, lower corn prices probably would stimulate increased corn utilization in other markets, but sales revenue would not be as high. The lower prices and sales revenue would be likely to result in higher federal spending on corn payments to farmers, as long as corn prices were below the price triggering federal loan deficiency subsidies.

ETHANOL REFINING AND PRODUCTION

According to the Renewable Fuels Association, about 55% of the corn used for ethanol is processed by "dry" milling plants (a grinding process) and the other 45% is processed by "wet" milling plants (a chemical extraction process). The basic steps of both processes are as follows. First, the corn is processed, with various enzymes added to separate fermentable sugars. Next, yeast is added to the mixture for fermentation to make alcohol. The alcohol is then distilled to fuel-grade ethanol that is 85-95% pure.[6] Finally, for fuel and industrial purposes the ethanol is denatured with a small amount of a displeasing or noxious chemical to make it unfit for human consumption.[7]

In the U.S. the denaturant for fuel ethanol is gasoline. Ethanol is produced largely in the Midwest corn belt, with almost 90% of production occurring in five states: Illinois, Iowa, Nebraska, Minnesota and Indiana. Because it is generally less expensive to produce ethanol close to the feedstock supply, it is not surprising that the top five corn-producing states in the U.S. are also the top five ethanol-producers. Most ethanol use is in the metropolitan centers of the Midwest, where it is produced. When ethanol is used in other regions, shipping costs tend to be high, since ethanol-blended gasoline cannot travel through petroleum pipelines. This geographic concentration is an obstacle to the use of ethanol on the East and West Coasts. The potential for expanding production geographically is a motivation behind research on ethanol, since if regions could locate production facilities closer to the point of consumption, the costs of using ethanol could be lessened. Furthermore, if regions could produce fuel ethanol from local crops, there would be an increase in regional agricultural income.

Ethanol production is also concentrated among a few large producers. The top five companies account for approximately 60% of production capacity, and the top ten companies account for approximately 75% of production capacity. (See Table 2.) Critics of the ethanol industry in general--and specifically of the ethanol tax incentives--argue that the tax incentives for ethanol production equate to "corporate welfare" for a few large producers.[8]

Overall, domestic ethanol production capacity is approximately 2.0 billion gallons per year. Consumption is expected to increase from 1.7 billion gallons per year in 2000 to approximately 2.6 billion gallons per year in 2005. Production will need to increase proportionally to meet the increased demand.[9] However, if the Clean Air Act is amended to limit or ban MTBE, ethanol production capacity may expand at a faster rate. This is especially true if MTBE is banned while maintaining the oxygenate requirements, since ethanol is the most likely substitute for MTBE.

Fuel is not the only output of an ethanol facility, however. Co-products play an important role in the profitability of a plant. In addition to the primary ethanol output, the corn wet milling generates corn gluten feed, corn gluten meal, and corn oil, and dry milling creates distillers grains. Corn oil is used as a vegetable oil and is higher priced than soybean oil.

Approximately 12 million metric tons of gluten feed, gluten meal, and dried distillers grains are produced in the United States and sold as livestock feed annually. A major market for corn gluten feed and meal is the European Union, which imported nearly 5 million metric tons of gluten feed and meal during FY1998.

Table 2. Top 10 Ethanol Producers by Capacity, 2000

Million Gallons	Per Year
Archer Daniels Midland (ADM)	797
Minnesota Corn Processors	110
Williams Energy Services	100
Cargill	100
New Energy Corp	85
Midwest Grain Products	78
High Plains Corporation	70
Chief Ethanol	62
AGP	52
A.E. Staley	45
Chief Ethanol	40
All Others	508
U.S. Total	2007

Source: Renewable Fuels Association, Ethanol Industry Outlook 2001.

Revenue from the ethanol byproducts help offset the cost of corn. The net cost of corn relative to the price of ethanol (the ethanol production margin) and the difference between ethanol and wholesale gasoline prices (the fuel blending margin) are the major determinants of the level of ethanol production. Currently, the ethanol production margin is high because of the low price of corn. At the same time, the wholesale price of gasoline is increasing against the price of ethanol, which encourages the use of ethanol as an octane enhancer.

FUEL CONSUMPTION

Approximately 1.4 billion gallons of ethanol fuel were consumed in the United States in 1999, mainly blended into E10 gasohol. While large, this figure represents only 1.2% of the approximately 125 billion gallons of gasoline consumption in the same year.[10] According to DOE, ethanol consumption is expected to grow to 2.6 billion gallons per year in 2005 and 3.3 billion gallons per year in 2020. This would increase ethanol's market share to approximately 1.5% by 2005. This 1.5% share is projected to remain constant through 2020 [11].

The most significant barrier to wider use of fuel ethanol is its cost. Even with tax incentives for ethanol producers (see the section on Economic Effects), the fuel tends to be more expensive than gasoline per gallon. Furthermore, since fuel ethanol has a somewhat lower energy content, more fuel is required to travel the same distance. This energy loss leads to an approximate 3% decrease in miles-per-gallon vehicle fuel economy with gasohol [12].

However, ethanol's chemical properties make it very useful for some applications, especially as an additive in gasoline. Major stimuli to the use of ethanol have been the oxygenate requirements of the Reformulated Gasoline (RFG) and Oxygenated Fuels programs of the Clean Air Act.[13] Oxygenates are used to promote more complete combustion of gasoline, which reduces carbon monoxide and volatile organic compound (VOC) emissions.[14] In addition, oxygenates can replace other chemicals in gasoline, such as benzene, a toxic air pollutant (see the section on Air Quality).

The two most common oxygenates are ethanol and methyl tertiary butyl ether (MTBE). MTBE, primarily made from natural gas or petroleum products, is preferred to ethanol in most regions because it is generally much less expensive, is easier to transport and distribute, and is available in greater supply. Because of different distribution systems and blending processes (with gasoline), substituting one oxygenate for another can lead to significant cost increases.

Despite the cost differential, there are several possible advantages of using ethanol over MTBE. Ethanol contains 35% oxygen by weight—twice the oxygen content of MTBE. Furthermore, since ethanol is produced from agricultural products, it has the potential to be a sustainable fuel, while MTBE is produced from natural gas and petroleum, fossil fuels. In addition, ethanol is readily biodegradable, eliminating some of the potential concerns about groundwater contamination that have surrounded MTBE (see the section on MTBE).

Both ethanol and MTBE also can be blended into otherwise nonoxygenated gasoline to raise the octane rating of the fuel, and therefore improve its combustion properties. High-performance engines and older engines often require higher octane fuel to prevent early ignition, or "engine knock." Other chemicals may be used for the same purpose, but some of these alternatives are highly toxic, and some are regulated as pollutants under the Clean Air Act.[15] Furthermore, since these additives do not contain oxygen, they do not result in the same emissions reductions as oxygenated gasoline.

In purer forms, ethanol can also be used as an alternative to gasoline in vehicles specifically designed for its use, although this only represents approximately 0.2% of ethanol consumption in the U.S. The federal government and state governments, along with businesses in the alternative fuel industry, are required to purchase alternative-fueled vehicles by the Energy Policy Act of 1992.[16] In addition, under the Clean Air Act Amendments of 1990, municipal fleets can use alternative fuel vehicles to mitigate air quality problems. Blends of 85% ethanol with 15% gasoline (E85), and 95% ethanol with 5% gasoline (E95) are currently considered alternative fuels by the Department of Energy.[17] The small amount of gasoline added to the alcohol helps prevent corrosion of engine parts, and aids ignition in cold weather.

Approximately 1.7 million gasoline-equivalent gallons (GEG)[18] of E85, and 59 thousand GEG of E95 were consumed in 1998, mostly in Midwestern states.[19] (See Table 3.) One reason for the relatively low consumption of E85 and E95 is that there are relatively few vehicles on the road that operate on these fuels. In 1998, approximately 13,000 vehicles were fueled by E85 or E95,[20] [21] as compared to approximately 210 million gasoline- and diesel-fueled vehicles that were on the road in the same year.[22] One obstacle to the use of alternative fuel vehicles is that they are generally more expensive than conventional vehicles, although this margin has decreased in recent years with newer technology. Another obstacle is that, as was stated above, fuel ethanol is generally more expensive than gasoline or diesel fuel. In addition, there are very few fueling sites for E85 and E95, especially outside of the Midwest.

Table 3. Estimated U. S. Consumption of Fuel Ethanol, MTBE and Gasoline (Thousand Gasoline-Equivalent Gallons)

1994	1996	1998	2000	(projected)
E85	80	694	1,727	3,283
E95	140	2,699	59[a]	59
Ethanol in Gasohol (E10)	845,900	660,200	916,000	908,700
MTBE in Gasoline	2,108,800	2,749,700	2,915,600	3,111,500
Gasoline[b]	113,144,000	117,783,000	122,849,000	127,568,000

Source: Department of Energy, Alternatives to Traditional Transportation Fuels1998.

[a]A major drop in E95 consumption occurred between 1997 and 1998 because of a significant decrease in the number of E95-fueled vehicles in operation (347 to 14), due to the elimination of an ethanol-fueled bus fleet in California.

[b.]Gasoline consumption includes ethanol in gasohol and MTBE in gasoline.

Research and Development in Cellulosic Feedstocks

For ethanol to play a more important role in U.S. fuel consumption, the fuel must become price-competitive with gasoline. Since a major part of the total production cost is the cost of feedstock, reducing feedstock costs could lead to lower wholesale ethanol costs. For this reason, there is a great deal of interest in the use of cellulosic feedstocks, which include low-value waste products, such as recycled paper, or dedicated fuel crops, such as switch grass. A dedicated fuel crop is one that would be grown and harvested solely for the purpose of fuel production.

However, as the name indicates, cellulosic feedstocks are high in cellulose, and cellulose cannot be fermented. Cellulose must first be broken down into simpler carbohydrates, and this can add an expensive step to the process. Therefore, research has focused on both reducing the process costs for cellulosic ethanol, and improving the availability of cellulosic feedstocks. On August 12, 1999, the Clinton Administration announced the Biobased Products and Bioenergy Initiative, which aims to triple the use of fuels and products derived from biomass by 2010.[23] Research and development covers all forms of biobased products, including lubricants, adhesives, building materials, and biofuels. Because federal research into cellulosic ethanol is ongoing, it is likely that funding would increase under the initiative.

Costs and Benefits of Fuel Ethanol

Economic Effects

Given that a major constraint on the use of ethanol as an alternative fuel, and as an oxygenate, is its high price, ethanol has not been competitive with gasoline as a fuel. Wholesale ethanol prices, before incentives from the federal government and state governments, are generally twice that of wholesale gasoline prices. With federal and state incentives, however, the effective price of ethanol is much lower. Furthermore, gasoline prices have risen recently, making ethanol more attractive.

The primary federal incentive to support the ethanol industry is the 5.4¢ per gallon exemption that blenders of gasohol (E10) receive from the 18.4¢ federal excise tax on motor fuels.[24] Because the exemption applies to blended fuel, of which ethanol comprises only 10%, the exemption provides for an effective subsidy of 54¢ per gallon of pure ethanol. (See Table 4.) It is argued that the ethanol industry could not survive without the tax exemption. An economic analysis conducted in 1998 by the Food and Agriculture Policy Research Institute, in conjunction with the congressional debate over extension of the tax exemption, concluded that ethanol production from corn would decline from 1.4 billion gallons per year, and stabilize at about 290 million gallons per year, if the exemption were eliminated.[25]

Table 4. Price of Pure Ethanol Relative to Gasoline

July 1998 to June 1999	
Ethanol Wholesale	103 ¢/gallon
PriceaAlcohol Fuel Tax	54 ¢/gallon
Incentive Effective Price of	49 ¢/gallon
Ethanol Gasoline Wholesale Priceb	46 ¢/gallon

Source: Hart's Oxy-Fuel News; Energy Information Agency, Petroleum Marketing Monthly.
[a]This is the average price for pure ("neat") ethanol.
[b]This is the average rack price for regular conventional gasoline (i.e. non-oxygenated, standard octane).

The tax exemption for ethanol is criticized by some as a corporate subsidy,[26] because, in this view, it encourages the inefficient use of agricultural and other resources, and deprives the Highway Trust Fund of needed revenues.[27] In 1997, the General Accounting Office estimated that the tax exemption would lead to approximately $10.4 billion in foregone Highway Trust Fund revenue over the 22 years from FY1979 to FY2000.[28] The petroleum industry opposes the incentive because it also results in reduced use of petroleum.

Proponents of the tax incentive argue that ethanol leads to better air quality, and that substantial benefits flow to the agriculture sector due to the increased demand for corn created by ethanol. Furthermore, they argue that the increased market for ethanol leads to a stronger U.S. trade balance, since a smaller U.S. ethanol industry would lead to increased imports of MTBE to meet the demand for oxygenates.[29]

Air Quality

One of the main motivations for ethanol use is improved air quality. Ethanol is primarily used in gasoline to meet minimum oxygenate requirements of two Clean Air Act programs. Reformulated gasoline (RFG)[30] is used to reduce vehicle emissions in areas that are in severe or extreme nonattainment of National Ambient Air Quality Standards (NAAQS) for ground-level ozone.[31] Ten metropolitan areas, including New York, Los Angeles, Chicago, Philadelphia, and Houston, are covered by this requirement, and many other areas with less severe ozone problems have opted into the program, as well. In these areas, RFG is used year-round. By contrast, the Oxygenated Fuels program operates only in the winter months in 20 areas[32] that are listed as carbon monoxide (CO) nonattainment areas.[33] EPA states that RFG has led to significant improvements in air quality, including a 17% reduction in volatile organic compounds (VOCs) emissions from vehicles, and a 30% reduction in toxic emissions. Furthermore, according to EPA "ambient monitoring data from the first year of the RFG program (1995) also showed strong signs that RFG is working. For example, detection of benzene (one of the air toxics controlled by RFG, and a known human carcinogen) declined dramatically, with a median reduction of 38% from the previous year."[34]

However, the need for oxygenates in RFG has been questioned. Although oxygenates lead to lower emissions of VOCs, and CO, they may lead to higher emissions of nitrogen oxides (NOX). Since all three contribute to the formation of ozone, the National Research Council recently concluded that while RFG certainly leads to improved air quality, the oxygenate requirement in RFG may have little overall impact on ozone formation.[35]

Furthermore, the high price of Midwest gasoline in Summer 2000 has raised further questions about the RFG program (see the section on Phase 2 Reformulated Gasoline). Evidence that the most widely-used oxygenate, methyl tertiary butyl ether (MTBE), contaminates groundwater has led to a push by some to eliminate the oxygen requirement in RFG. MTBE has been identified as an animal carcinogen, and there is concern that it is a possible human carcinogen. In California, MTBE will be banned as of December 31, 2002, and the state is lobbying Congress for a waiver to the oxygen requirement (see section on MTBE). Other states, such as states in the Northeast, are also seeking waivers.

If the oxygenate requirements were eliminated, some refiners claim that the environmental goals of the RFG program could be achieved through cleaner, although potentially more costly, gasoline that does not contain any oxygenates.[36] These claims have added to the push to remove the oxygen requirement and allow refiners to produce RFG in the most cost-effective manner, whether or not that includes the use oxygenates. However, some environmental groups are concerned that an elimination of the oxygenate requirements would compromise air quality gains resulting from the current standards, since oxygenates also displace other harmful chemicals in gasoline. This potential for "backsliding" is a result of the fact that the current performance of RFG is substantially better that the Clean Air Act requires. If the oxygenate standard were eliminated, environmental groups fear that refiners would only meet the requirements of the law, as opposed to maintaining the current overcompliance.

While the potential ozone benefit from oxygenates in RFG has been questioned, there is little dispute that the winter Oxy-Fuels program has led to lower emissions of CO. The Oxy-Fuels program requires oxygenated gasoline in the winter months to control CO pollution in NAAQS nonattainment areas for the CO standard. However, this program is small relative to the RFG program.[37]

The air quality benefits from purer forms of ethanol can also be substantial. Compared to gasoline, use of E85 and E95 can result in a 30-50% reduction in ozone-forming emissions. And while the use of ethanol also leads to increased emissions of acetaldehyde, a toxic air pollutant, as defined by the Clean Air Act, these emissions can be controlled through the use of advanced catalytic converters.[38] However, as was stated above, these purer forms of ethanol have not seen wide use.

Climate Change

Another potential environmental benefit from ethanol is the fact that it is a renewable fuel. Proponents of ethanol argue that over the entire fuelcycle [39] it has the potential to reduce greenhouse gas emissions from automobiles relative to gasoline, therefore reducing the risk of possible global warming.

Because ethanol (C2H5OH) contains carbon, combustion of the fuel necessarily results in emissions of carbon dioxide (CO2), the primary greenhouse gas. However, since photosynthesis (the process by which plants convert light into chemical energy) requires absorption of CO2, the growth cycle of the feedstock crop can serve--to some extent--as a "sink" that absorbs some of these emissions. In addition to CO2 emissions, the emissions of other greenhouse gases may increase or decrease depending on the fuel cycle.[40]

According to Argonne National Laboratory, using E10, vehicle greenhouse gas emissions (measured in grams per mile) are approximately 1% lower than with the same vehicle using gasoline. With improvements in production processes, by 2010, the reduction in greenhouse gas emissions from ethanol relative to gasoline could be as high as 8-10% for E10, while the use of E95 could lead to significantly higher reductions.[41]

While some studies have called into question the efficiency of the ethanol production process, most recent studies find a net energy gain.[42] If true, then the overall reductions in greenhouse gas emissions would be diminished, due to higher fuel consumption during the production process.

Energy Security

Another frequent argument for the use of ethanol as a motor fuel is that it reduces U.S. reliance on oil imports, making the U.S. less vulnerable to a fuel embargo of the sort that occurred in the 1970s, which was the event that initially stimulated development of the ethanol industry. According to Argonne National Laboratory, with current technology the use of E10 leads to a 3% reduction in fossil energy use per vehicle mile, while use of E95 could lead to a 44% reduction in fossil energy use.[43]

However, other studies contradict the Argonne study, suggesting that the amount of energy needed to produce ethanol is roughly equal to the amount of energy obtained from its combustion, which could lead to little or no reductions in fossil energy use.[44] Thus, if the energy used in ethanol production is petroleum-based, ethanol would do nothing to contribute to energy security. Furthermore, as was stated above, fuel ethanol only displaces approximately 1.2% of gasoline consumption in the United States.

This small market share led GAO to conclude that the ethanol tax incentive has done little to promote energy security.[45] Furthermore, since ethanol is currently dependent on the U.S. corn supply, any threats to this supply (e.g. drought), or increases in corn prices, would negatively affect the cost and/or supply of ethanol. This happened when high corn prices caused by strong export demand in 1995 contributed to an 18% decline in ethanol production between 1995 and 1996.

POLICY CONCERNS AND CONGRESSIONAL ACTIVITY

Recent congressional interest in ethanol fuels has mainly focused on three sets of issues: 1) implementation of Phase 2 of the RFG program; 2) a possible phase-out of MTBE; and 3) the alcohol fuel tax incentives.

Phase 2. Reformulated Gasoline

Under the new Phase 2 requirements of the RFG program, which took effect in 2000, gasoline sold in the summer months (beginning June 1) must meet a tighter volatility standard.[46] Reid Vapor Pressure (RVP) is a measure of volatility, with higher numbers indicating higher volatility. Because of its physical properties, ethanol has a higher RVP than MTBE. Therefore, to make Phase 2 RFG with ethanol, the gasoline, called RBOB,[47] must have a lower RVP. This low-RVP fuel is more expensive to produce, leading to higher production costs for ethanol-blended RFG. Before the start of Phase 2, estimates of the increased cost to produce RBOB for ethanol-blended RFG ranged from 2 to 4 cents per gallon, to as much as 5 to 8 cents per gallon.[48].

In Summer 2000, RFG prices in Chicago and Milwaukee were considerably higher than RFG prices in other areas, and it has been argued that the higher production cost for RBOB was one cause. However, not all of the price difference is attributable to the new Phase 2 requirements or the use of ethanol. Conventional gasoline prices in the Midwest were also high compared with gasoline prices in other areas. High crude oil prices, low gasoline inventories, pipeline problems, and uncertainties over a patent dispute pushed up prices for all gasoline in the Midwest.

To decrease the potential for price spikes, on March 15, 2001, EPA announced that Chicago and Milwaukee will be allowed to blend slightly higher RVP reformulated gasoline during the summer months.[49] This action is not a change in regulations but a revision of EPA's enforcement guidelines. In addition to EPA's action, one possible regulatory option that has been suggested to control summer RFG prices is a more significant increase in the allowable RVP under Phase 2. Although the volatility standard is set by the Clean Air Act, the Environmental Protection Agency (EPA) is currently reviewing whether credits from ethanol's improved performance on carbon monoxide emissions are possible as an offset to its higher volatility. Legislative options have included eliminating the oxygenate standard for RFG, or suspending the program entirely. However, some in the petroleum industry suggest that additional changes to fuel requirements could further disrupt gasoline supplies. No bills to address the RVP issue have been introduced in the 107th Congress.

MTBE

Another key issue involving ethanol is the current debate over MTBE. Since MTBE, a possible human carcinogen, has been found in groundwater in some states (especially in California), there has been a push both in California and nationally to ban MTBE.[50] In March 1999, California's Governor Davis issued an Executive Order requiring that MTBE be phased out of gasoline in the state by December 31, 2002. Arizona, Connecticut, Iowa, Minnesota, Nebraska, New York, and South Dakota have also instituted limits or bans on MTBE. In July 1999, an advisory panel to EPA recommended that MTBE use should be "reduced substantially."[51]

A possible ban on MTBE could have serious consequences for fuel markets, especially if the oxygenate requirements remain in place. Since ethanol is the second most used oxygenate, it is likely that it would be used to replace MTBE. However, there is not currently enough U.S. production capacity to meet the potential demand. Therefore, it would likely be necessary to phase out MTBE over time, as opposed to an immediate ban.

Furthermore, the consumer price for oxygenated fuels would likely increase because ethanol, unlike MTBE, cannot be shipped through pipelines and must be mixed close to the point of sale, adding to delivery costs. Increased demand for oxygenates could also be met through imports from countries such as Brazil, which is a leader worldwide in fuel ethanol production, and currently has a surplus.[52]

While a ban on MTBE would seem to have positive implications for ethanol producers, it could actually work against them. Because MTBE is more commonly used in RFG and high-octane gasoline, and because current ethanol production can not currently meet total U.S. demand for oxygenates and octane, there is also a push to suspend the oxygenate requirement in RFG, which would remove a major stimulus to the use of fuel ethanol. Furthermore, environmental groups and state air quality officials, although supportive of a ban on MTBE, are concerned over the possibility of "backsliding" if the oxygenate standard is eliminated. Because current RFG formulations have a lower level of toxic substances than is required under the Clean Air Act, there are concerns that new RFG formulations without oxygenates will meet the existing standard, but not the current level of overcompliance.

On March 20, 2000, the Clinton Administration announced a plan to reduce or eliminate MTBE use, and to promote the use of ethanol. Although no legislative language was suggested, the framework included three recommendations. The first was to "provide the authority to significantly reduce or eliminate the use of MTBE." The second recommendation was that "Congress must ensure that air quality gains are not diminished." The third was that "Congress should replace the existing oxygenate requirement in the Clean Air Act with a renewable fuel standard for all gasoline." Moreover, the Clinton Administration discussed the possibility of limiting the use of MTBE through the Toxic Substances Control Act (P.L. 94-469), which gives EPA the authority to control any substance that poses unreasonable risk to health or the environment. However, this process could take several years.[53]

MTBE producers argued that such an initiative will lower clean air standards, and raise gasoline prices, while ethanol producers and some environmental groups were generally supportive of the announcement.[54] In the 107th Congress, six MTBE-related bills have been introduced. (See Appendix 1.) All have been referred to Committee. These bills address different facets of the MTBE issue, including limiting or banning the use of MTBE, granting waivers to the oxygenate requirement, and authorizing funding for MTBE cleanup.

Alcohol Fuel Tax Incentives[55]

As stated above, the exemption that ethanol-blended fuels receive from the excise tax on motor fuels is controversial. The incentive allows fuel ethanol to compete with other additives, since the wholesale price of ethanol is so high. Proponents of ethanol argue that this exemption lowers dependence on foreign imports, promotes air quality, and benefits farmers.[56] A related, albeit smaller incentive for ethanol production is the small ethanol producers tax credit. This credit provides 10 cents per gallon for up to 15 million gallons of annual production by a small producer.[57] Opponents of the tax incentives argue that the incentives promote an industry that could not exist on its own, and reduce potential fuel tax revenue. Despite objections from opponents, Congress in 1998 extended the motor fuels tax exemption through 2007, but at slightly lower rates (P.L. 105-178). A bill in the 107th Congress, S. 312, would increase the size of a covered producer under the small producer tax credit. (See Appendix 1.)

CONCLUSION

As a result of the current debate over the future of MTBE in RFG, and the RFG program in general, the future of the U.S. ethanol industry is uncertain. A ban on MTBE would greatly expand the market for ethanol, while an elimination of the oxygenate requirement would remove a major stimulus for its use. Any changes in the demand for ethanol will have major effects on corn producers, who rely on the industry as a partial market for their products.

The current size of the ethanol industry is depends significantly on federal laws and regulations that promote its use for air quality and energy security purposes, as well as tax incentives that lessen its cost to consumers. Without these, it is likely that the industry would shrink substantially in the near future. However, if fuel ethanol process costs can be decreased, or if gasoline prices increase, ethanol could increase its role in U.S. fuel consumption.

REFERENCES

[1] Industrial uses include perfumes, aftershaves, and cleansers.

[2] U.S. Department of Energy (DOE), Energy Information Administration (EIA). *Alternatives to Traditional Transportation Fuels* 1998. October 1999.

[3] One bushel of corn generates approximately 2.5 gallons of ethanol.

[4] Utilization data are used, rather than production, due to the existence of carryover stocks. Corn utilization data address the total amount of corn used within a given period.

[5] Detailed explanations are available in *CRS Report* RS20271, Grain, Cotton, and Oilseeds: Federal Commodity Support, and *CRS* 98-744, Agricultural Marketing Assistance Loans and Loan Deficiency Payments.

[6] The byproduct of the dry milling process is distillers dried grains. The byproducts of wet milling are corn gluten feed, corn gluten meal, and corn oil. Distillers dried grains, corn gluten feed, and corn gluten meal are used as livestock feed.

[7] Renewable Fuels Association, *Ethanol Industry Outlook* 2001, Clean Air, Clean Water, Clean Fuel.

[8] James Bovard, Archer Daniels Midland: A Case Study in Corporate Welfare. *Cato Institute*. September 26, 1995.

[9] DOE, EIA, *Annual Energy Outlook* 2001. December 22, 2000. Table 18.

[10] DOE, EIA, Alternatives to Traditional Transportation Fuels 1998. October 1999. Table 10.

[11] DOE, EIA, *Annual Energy Outlook* 2001. December 22, 2000. Tables 2 and 18.

[12] It should be noted that the use of ethanol does not effect the efficiency of an engine. There is simply less energy in one gallon of ethanol than in one gallon of gasoline. Fuel Ethanol: Background and Public Policy Issues 95.

[13] Section 211, subsections k and m (respectively). 42 U.S.C. 7545.

[14] CO, VOCs and nitrogen oxides (NOX)are the main precursors to ground-level ozone.

[15] Lead was commonly used as an octane enhancer until it was phasedout through the mid-1980s (lead in gasoline was completely banned in 1995), due to the fact that it disables emissions control devices, and because it is toxic to humans.

[16] P.L. 102-486.

[17] More diluted blends of ethanol, such as E10, are considered to be "extenders" of gasoline, as opposed to alternatives.

[18] Since different fuels produce different amounts of energy per gallon when consumed, the unit of a gasoline-equivalent gallon (GEG) is used to compare total energy consumption.

[19] DOE, EIA, *Alternatives to Traditional Transportation Fuels* 1998.

[20] Ibid.

[21] In 1997, some manufacturers made flexible E85/gasoline fueling capability standard on some models. It is expected, however, that most of these vehicles will be fueled by gasoline.

[22] Stacy C. Davis, DOE, *Transportation Energy Data Book:* Edition 20. November 2000.

[23] Executive Order 13134. August 12, 1999.

[24] 26 U.S.C. 40.

[25] *Food and Agriculture Policy Research Institute*. Effects on Agriculture of Elimination of the Excise Tax Exemption for Fuel Ethanol, Working Paper 01-97, April 8, 1997.

[26] James Bovard. p. 8.

[27] *U.S. General Accounting Office, Effects of the Alcohol Fuels Tax Incentives. March, 1997.*

[28] Ibid.

[29] Katrin Olson, "USDA Shows Losses Associated with Eliminating Ethanol Incentive," *Oxy-Fuel News*. May 19, 1997. p. 3.

[30] *Clean Air Act*, Section 211, subsection k. 42 U.S.C. 7545.

[31] Ground-level ozone is an air pollutant that causes smog, adversely affects health, and injures plants. It should not be confused with stratospheric ozone, which is a natural layer some 6 to 20 miles above the earth and provides a degree of protection from harmful radiation.

[32] Only the Los Angeles and New York areas are subject to both programs.

[33] *Clean Air Act*, Section 211, subsection m. 42 U.S.C. 7545. Brent D. Yacobucci and Jasper 96 er Womach

[34] Margo T. Oge, Director, *Office of Mobile Sources*, U.S. EPA, Testimony Before the Subcommittee on Energy and Environment of the Committee on Science, U.S. House of Representatives. September 14, 1999.
[35] National Research Council, Ozone-Forming Potential of Reformulated Gasoline. May, 1999.
[36] Al Jessel, Senior Fuels Regulatory Specialist of Chevron Products Company, Testimony Before the House Science Committee Subcommittee on Energy and Environment. September 30, 1999.
[37] In 1998, an average of 90.9 million gallons per day of RFG were sold in the U.S., as opposed to 8.0 million gallons per day of Oxy-Fuel gasoline.
[38] California Energy Commission, Ethanol-Powered Vehicles.
[39] The fuel-cycle consists of all inputs and processes involved in the development, delivery and final use of the fuel.
[40] For example, nitrous oxide emissions tend to increase with ethanol use because nitrogen-based fertilizers are used extensively in agricultural production.
[41] M. Wang, C. Saricks, and D. Santini, "Effects of Fuel Ethanol on Fuel-Cycle Energy and Greenhouse Gas Emissions." Argonne National Laboratory.
[42] Hosein Shapouri, James A. Duffield, and Michael S. Graboski, USDA, Economic Researc Service, Estimating the Net Energy Balance of Corn Ethanol. July 1995.
[43] Wang, et. al. p. 1
[44] Shapouri, et. al. Table 1.
[45] U.S. General Accounting Office, Effects of the Alcohol Fuels Tax Incentives. March, 1997.
[46] Volatility of gasoline is its tendency to evaporate.
[47] RBOB: Reformulated Gasoline Blendstock for Oxygenate Blending.
[48] Estimates from the Renewable Fuels Association and EPA, respectively.
[49] Pamela Najer, "Refiners Get Flexibility to Blend Ethanol for Summer Fuel Supply in Two Cities," Daily Environment Report. March 19, 2001. p. A9.
[50] For more information, see CRS Report 98-290 ENR, MTBE in Gasoline: Clean Air and Drinking Water Issues.
[51] Blue Ribbon Panel on Oxygenates in Gasoline, Achieving Clean Air and Clean Water: The Report of the Blue Ribbon Panel on Oxygenates in Gasoline.
[52] Adrian Schofield, "Brazilian Ambassador Sees Opportunity in United States Ethanol Market," New Fuels & Vehicles Report. September 16, 1999. p. 1.
[53] U.S. Environmental Protection Agency, Headquarters Press Release: Clinton-Gore Administration Acts to Eliminate MTBE, Boost Ethanol. March 20, 2000.
[54] Jim Kennett, "Government Seeks to Ban Gas Additive," Houston Chronicle. March 21, 2000. p. A1.
[55] For more information, see CRS Report 98-435E, Alcohol Fuels Tax Incentives.
[56] U.S. General Accounting Office (GAO), Effects of the Alcohol Fuels Tax Incentives. March, 1997.
[57] Defined as having a production capacity of less than 30 million gallons per year.

INDEX